American Farms,
American Food

American Farms, American Food

A Geography of Agriculture and Food Production in the United States

John C. Hudson and
Christopher R. Laingen

LEXINGTON BOOKS
Lanham • Boulder • New York • London

Published by Lexington Books
An imprint of The Rowman & Littlefield Publishing Group, Inc.
4501 Forbes Boulevard, Suite 200, Lanham, Maryland 20706
www.rowman.com

Unit A, Whitacre Mews, 26-34 Stannary Street, London SE11 4AB

British Library Cataloguing in Publication Information Available

Library of Congress Cataloging-in-Publication Data
Names: Hudson, John C., author. | Laingen, Christopher R., author.
Title: American farms, American food : a geography of agriculture and food production
    in the United States / John C. Hudson and Christopher R. Laingen.
Description: Lanham, Maryland : Lexington Books [2016] | Includes bibliographical
    references.
Identifiers: LCCN 2016023405 (print) | LCCN 2016025637 (ebook) |
    ISBN 9781498508209 (cloth : alk. paper) | ISBN 9781498508216 (Electronic)
Subjects: LCSH: Agriculture—United States. | Farms—United States. | Agricultural
    productivity—United States. | Food supply—United States.
Classification: LCC S441 .H83 2016 (print) | LCC S441 (ebook) | DDC 630.973—dc23
LC record available at https://lccn.loc.gov/2016023405

♾️™ The paper used in this publication meets the minimum requirements of American
National Standard for Information Sciences—Permanence of Paper for Printed Library
Materials, ANSI/NISO Z39.48-1992.

Printed in the United States of America

Contents

Preface

This book is based on the authors' shared conviction that the world needs a better understanding of how agriculture and food are connected. Anyone who has ever tried to search online for either of these subjects knows there is no lack of information about food and agriculture. Food is the consumption side. Websites about food, food safety, food security, and environmental sustainability are numerous, information-packed, and easily accessed. Agriculture represents the production side. Information about agriculture and farming is available, but it is difficult for the non-expert to access it, and there seems to be little interest in doing so on the part of the general public. What is produced on farms ends up as our food, yet the connections are not always clear. This book's task is to shed light on the matter by treating farm production not only for its own sake but also to show what happens after the products leave the farm.

The U.S. Department of Agriculture (USDA) is the federal agency whose data, analyses, and published reports bridge the divide between agriculture and food studies. We have relied heavily on USDA materials in writing this book and in making the maps that portray the geography of food and agriculture. Our allocation of subject matter into chapters follows USDA orthodoxy as well, with a basic division between crops and livestock and then subdivision within each of those to focus on specific crops or types of livestock.

Chapter 1: Farms and Food begins the book with a discussion of farm direct sales—what consumers buy directly from farmers—which is followed by an overview of agricultural production at the national scale. Chapter 2: The Family Farm shows the overwhelming role family-owned and -operated farms play in American agriculture and analyzes patterns of farm ownership and farm size. Chapter 3: The Corn Belt; and Chapter 4: Wheat and Grains cover the two major grain crops grown in this country and explain how their respective geographies have evolved over time. Other field crops, such as soybeans, barley, and rice, are included in these chapters as well.

The three chapters that follow the grain crops treat each of the livestock and animal product sectors individually. Chapter 5: Dairy; Chapter 6: Pork and Beef; and Chapter 7: Poultry detail the history, geography, and recent controversies surrounding animal production on farms and the role of the meat industry. Chapter 8: Fruits and Vegetables covers dozens of specialty crops in these sectors, describing how and where the crops are grown and the role that marketing and trade play in getting these products into the consumers' hands. Chapter 9: Organic Farms and Organic Food develops the history of organic production and certification and examines the expanding role it plays in the American food supply. Chapter 10: The Reserved Lands details the federal programs that, since the 1930s, have held some lands out of production mainly for conservation purposes. The book ends with a data-oriented appendix entitled Keeping Track of Production, which is a guide to print and electronic data sources about American agriculture for those interested in researching some of the subjects covered in this book.

We have benefited from the discussions we have had on these subjects with other geographers with whom we have collaborated in presenting research papers at annual meetings of the Association of American Geographers in recent years. We would like to thank Roger Auch, Holly Barcus, Ryan Baxter, Michelle Bouchard, John Cross, Dawn Drake, John Fraser Hart, Darrell Napton, Ryan Reker, Brad Rundquist, and Susy Ziegler for their insights and comments on our work.

By far our greatest debt is to John Fraser Hart, who basically invented this subject and has generously given us his guidance and wise counsel over the years.

Chapter 1

Farms and Food

There is no doubting the connection between the food we eat and the farms from which it comes. Yet how many people have an idea how the two are connected? Many years ago most Americans lived on farms and they were intimately familiar with where their food came from because they produced it themselves. It took the United States more than a century to go from being a nation of food producers to being primarily one of food consumers. By 1940 every U.S. farmer was feeding an average of 19 people. Today that ratio has become one farmer feeding 155 people and the trend continues (USDA, Economic Research Service 2016b).

The productivity of American farms is a continuing success story that is based in part on technological advances in production and marketing that have kept pace with the times. Measured by the value of output, the person-hours required for labor, the acreage devoted to production, or the number of people fed by one farm family, American farms have steadily advanced in their efficiency and output for well over two centuries.

Early twentieth-century agriculture was labor-intensive, and it took place on a large number of small, diversified farms in rural areas where more than half of the U.S. population lived. These conditions have changed steadily over time. The 22 million work animals needed for agriculture in the early twentieth century have been replaced by 5 million tractors. The agricultural sector of the twenty-first century is concentrated on a small number of large, specialized farms in rural areas where less than one-fourth of the U.S. population lives (Dimitri and others 2005).

The drive toward increased efficiency on the farm has been relentless, but people's ties to farm life have lagged somewhat behind the pace of technological change itself. As late as 1958, 24% of the American population aged 18 and over—nearly 26 million adults—had been born on farms (Beale and others 1964). The farms on which they were born and reared were more diversified than the typical farm of today. Those who knew farm life two generations ago experienced at least some of the work and knew the satisfaction of growing one's own food.

This was a stage of transition that could not last for long, however. More than 16 million of those 26 million farm-born people of 1958 had already left the farm and were living in cities or small towns across the country. The next generation of the farm-born could be only a fraction of the size of the previous one. In today's U.S. population of 320 million even those with a farm-born parent or grandparent are in a small minority. Although the statistic is no longer officially tabulated, the number of Americans living on farms now hovers around 3 million.

Fewer Americans than ever before know farms from personal experience. But the lack of direct experience does not mean a lack of interest in what farmers produce or in how they produce it. Paralleling the technological changes in agriculture that have led to production efficiencies has been an increased consumer awareness and concern for how food is grown and marketed. This has come in part from a general increase in concern for one's own nutrition and diet but it also has come from the larger food processing companies that seek consumer satisfaction with the products they offer for sale. Consumer influence on agricultural production comes both directly from consumers and indirectly from the food processors who seek to predict the choices consumers will make.

Companies that process food intended for restaurants and supermarkets have played an increasing role in food production. The instrument that is most often used to monitor and control food supply is the production contract. Farmers who sign a contract to deliver a product on a given date at an agreed-upon price are shielded from some of the inherent risks that increase farm production costs. Food processing and packing companies know that consumers want a specific product and that they expect it to be available for purchase. The demand for differentiated agricultural products to meet specific consumer preferences has led to a growing popularity of agricultural contracts (MacDonald and others 2004).

Virtually all poultry produced in the United States today are raised and sold on a contract basis More than 90% of the cattle, hogs, and dairy products are produced on contract, as are more than half of the crops. Overall more than two-thirds of farm output in the United States is sold on contracts between farmers and processing companies (MacDonald 2015). The use of contractual relationships makes it possible to trace production of a food item directly back to the farm or group of farms where it originated. Product recalls and other safety checks obviously are made easier by this arrangement which promotes accountability. Control over what gets marketed is enhanced by the presence of contracts and the use of digital communications technology.

This approach is not what some people have in mind when they express a desire to have more control over their food, however. The rise in popularity of locally produced foods over the past two decades can be interpreted in part as a reaction against the removal of farms from one's own life experience. If food merely comes from someplace else, and if it arrives already packaged like other consumer goods, there seem to be steps missing in the process.

DIRECT SALES FROM FARMS

A small but growing fraction of what is produced on American farms is sold directly to consumers (Ekenem and others 2016). Known as farm direct sales, or direct-to-consumer sales, food that leaves the farm in this manner was a $1.3 billion business in the United States in 2012, more than triple the value recorded in 1992 (NASS 2012). Another category of farm sales involves food distributors, brokers, and aggregators who operate through dozens of "food hubs" that are scattered around the country. Intermediaries receive products from farms and sell to restaurants, grocery stores, and supermarkets. Their activities are not recorded in the Census of Agriculture but their sales amounted to $4.8 billion in 2012 (Vogel and Low 2015). Other direct sales are included in community-supported agriculture offerings (see Chapter 9). The truly direct sales are those made through periodic farmers markets where growers offer their own produce for sale (Martinez and others 2016).

Surveys conducted by a supermarket industry association have reported that 25% of grocery shoppers look for foods that are locally grown or produced (Food Marketing Institute 2015). Other reasons consumers give for favoring local food is the importance attached to environmental sustainability, which small-scale, local food producers are thought to enhance (Burnett and others 2011). Some studies have suggested that food offered in farmers markets is sold at cheaper prices compared with those charged by supermarkets. Price differences tend to vary as much by region, season of the year, and commodity as they do between farmers markets and other retail outlets, however, and the differences have been difficult to document (Low and others 2015).

The farms that make direct sales to consumers are typically "local" (say, within 100 miles of the markets they choose to serve) but they are not necessarily "small." There is a high degree of concentration in terms of farm size among farms making direct sales. Farms with gross annual incomes below $75,000 constitute 85% of the direct-sales farms but they make just 13% of direct food sales. At the other extreme, direct-sales farms grossing more than $350,000 in total sales make up just 5% of the farms but account for two-thirds of the sales (Low and others 2015, 11). It is a familiar story in modern agriculture, with the smaller number of large farms making up the larger share of production.

Table 1.1   Value of Farm Direct Sales Within 100 Miles of Selected Cities, 2012 (millions of dollars)

Worcester, MA	129.1	Detroit, MI	31.4
Baltimore, MD	120.3	Spartanburg, SC	29.4
Washington, DC	108.0	St. Cloud, MN	28.9
Philadelphia, PA	104.4	Los Angeles, CA	27.4
Rutland, VT	92.5	Raleigh, NC	25.4
Fresno, CA	75.6	Atlanta, GA	11.6
Oakland, CA	69.8	Tampa, FL	9.5
Syracuse, NY	46.7	Denver, CO	8.8
Portland, OR	44.1	Houston, TX	5.4
Milwaukee, WI	43.8	Birmingham, AL	5.3
Chicago, IL	43.5	Phoenix, AZ	4.8

Source: USDA Census of Agriculture

Farm-to-consumer sales are also concentrated geographically. Setting an arbitrary radius of 100 miles from the market as satisfying the local criterion, it is apparent that cities of the urbanized Northeast are comparatively well supported by direct sales (Table 1.1). Southern New England is especially well supplied, as are Washington DC, Philadelphia, and Baltimore. A slightly lower level of production takes place around Chicago, Milwaukee, and Detroit, although the Midwest region as a whole has a smaller population and presumably would also consume less produce. California, where local produce markets can operate all year, shows fewer direct sales coming from its farms than is true in southern New England but slightly more than the Midwest.

Farmers who make direct sales must scale their marketing efforts to the size of the local population. Large cities and their suburbs would naturally be a more attractive market than places where potential customers are limited to a scatter of small cities. But population is not always a good predictor of direct sales. Atlanta, Tampa, Denver, Houston, and Phoenix are large cities that may have many potential customers for direct sales, but farms in their vicinities make comparatively few such sales. In comparison, Spartanburg, South Carolina, and Raleigh, North Carolina, both much smaller cities, have a relatively high availability of locally produced food.

The national map of farm direct sales shows substantial place-to-place variation in local food availability (Figure 1.1). Since farmers markets necessarily depend on direct sales to supply the food they offer, the map shows that locally produced food is either unavailable or is in short supply in much of the country. The South is poorly served compared with the Midwest, Northeast, and Pacific Coast regions. The Great Plains, from the Dakotas and Montana to Texas, produces massive amounts of food for the American consumer, but practically none of it is marketed locally, no doubt in part because of the region's small population and comparative lack of cities. Hot summer

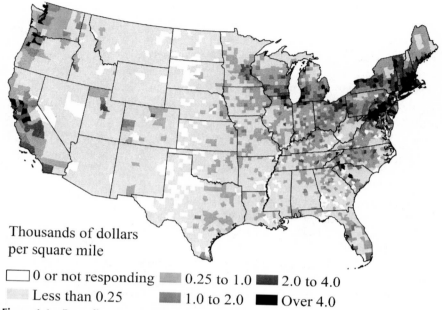

Thousands of dollars
per square mile

☐ 0 or not responding   ▨ 0.25 to 1.0   ■ 2.0 to 4.0
   Less than 0.25   ■ 1.0 to 2.0   ■ Over 4.0

Figure 1.1   Farm direct sales, 2012. *Source*: Created by authors, data from USDA Census of Agriculture 2012.

temperatures in places like Atlanta, Tampa, Houston, Phoenix, and Denver might diminish interest in shopping in what are generally little more than makeshift open-air sales venues, adding another reason for an apparent lack of interest in farm direct sales in the Sun Belt.

Farm direct sales make up a tiny fraction of the total sales from U.S. farms. The $1.31 billion in sales recorded in the 2012 Census amounted to just 0.33% of total U.S. farm sales, and even with the addition of food hub transactions—mostly with restaurants and retail stores—local food amounts to only 1.5% of U.S. total farm sales (Vogel and Low 2015).

VALUE OF FARM SALES

The $394.8 billion total sales from U.S. farms in 2012 was divided rather evenly between crops (53.8%) and livestock (46.1%), the latter including livestock products and poultry. Farms themselves are classified in terms of their largest source of income. In 2012 a total of 1,054,987 farms received the largest share of their income from crops, and a nearly identical 1,054,316 received most of their income from animal production.

Americans generally think of farms as being a mixture of crop and livestock production activities, and this is still true of some farms. But in 2012 the animal production farms received 93.4% of their income from the sales of animals or animal products. Crop farms derived 95.6% of their income from the sale of crops. The long-term trend toward specialization in either animals or crops thus continues, with most farms focusing on one or two crops or one type of animal.

As crop-producing farms and animal-producing farms are nearly equal in total number, the two sectors each require about the same amount of land (454 million acres in crop farms, 460 million in animal production). Agriculture as a whole has a widespread dispersion across the United States. The United States. has favorable climates, fertile soils, and smooth topography that few other large countries can match. Looking at the roughly 3,100 counties of the United States. in rank order, the least productive one thousand of them contribute just 3.9% of the value of farm sales. At the other end, the top 100 counties contributed about 26% of sales. While this is by no means an even distribution over the counties, it tells of a pattern of widespread production. It takes 500 counties to account for as much as two-thirds of U.S. farm production value (Figure 1.2).

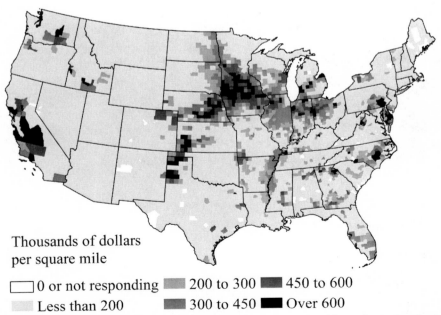

Thousands of dollars
per square mile

☐ 0 or not responding	▨ 200 to 300	▨ 450 to 600
▨ Less than 200	▨ 300 to 450	■ Over 600

Figure 1.2   Value of all sales from farms, 2012. *Source*: Created by authors, data from USDA Census of Agriculture 2012.

U.S. agriculture is widespread in total but it is regionally concentrated when viewed in terms of agricultural specialties. Thirty counties produced more than $1 billion worth of agricultural products each in 2012, and nine of the top ten of those were in California. By about any criterion that could be selected, California is the nation's leading agricultural state. It is the leading dairy producer (Chapter 5), contains 30% of the nation's vegetable acreage and 60% of the acreage in fruits and nuts (Chapter 8), and is the leading producer of organic food (Chapter 9).

Among the other counties producing more than a billion dollars in agricultural revenue were eight in Colorado, Kansas, and Texas which rank among the leading producers of beef cattle fattened for market (Chapter 6). Two more of the top thirty are in North Carolina and they produce hogs and turkeys. Clusters of intensive agriculture are scattered across the South, from Delaware to Texas, most of which are intensive areas of poultry production (Chapter 7).

Counties of the Corn Belt also rank high in production value (Chapter 3). Although the individual farms are specialized, most counties in this region that stretches from Ohio to the Dakotas to Nebraska produce large crops of corn and soybeans as well as thousands of hogs and beef cattle (Chapters 3 and 6). Wheat, a less valuable crop, dominates many counties in the Great Plains and Pacific Northwest (Chapter 4). But wheat produces less income per acre than most other specialties and the top wheat-producing counties are not distinct on the map of value of farm sales.

The $394.8 billion in U.S. farm sales cannot be equated with America's food budget, nor do the two even come close to being the same thing. In an average year, half of the wheat grown in the United States is exported. Half of the soybean crop is sold to other countries, and most of the rest goes for poultry and other animal feed. At present, 38% of the corn crop goes into fuel alcohol production, 38% becomes livestock feed, and 14% is exported (see Figure 3.5). The feed uses of these crops support the cattle, hogs, and poultry that Americans and others consume.

The United States is also an importer of agricultural commodities. Eleven million tons of grain, 12 million tons of fresh fruit, and 9 million tons of vegetables were imported from other countries in 2014 (USDA, Economic Research Service 2016a). The increase in purchasing power of economies around the globe has strengthened international trade and the United States is firmly a part of this trade.

Not everything grown on American farms goes for food, and not everything Americans eat is produced on American farms, but the overwhelmingly most important role of farms in this country is to provide food for the American people. "The Farm" of today, as in the past, is overwhelmingly a business that is owned by the family which lives there and operates it.

Chapter 2 focuses on the family farm—how it functions and how it has changed over time.

REFERENCES

Beale, C. L., J. C. Hudson, and V. J. Banks. 1964. *Characteristics of the U.S. Population by Farm and Nonfarm Origin.* Agricultural Economic Report No. 66. U.S. Dept. of Agriculture, Economic Research Service.

Burnett, P., T. H. Kuethe, and C. Price. 2011. Consumer Preference for Locally Grown Produce: An Analysis of Willingness-to-pay and Geographic Scale," *Journal of Agriculture, Food Systems and Community Development* 2(1): 269–278.

Dimitri, C., A. Effland, and N. Conklin. 2005. *The 20th Century Transformation of U.S. Agriculture and Farm Policy.* U.S. Dept. of Agriculture, Economic Research Service, Economic Information Bulletin, No. 3.

Ekenem, E., M. Mafuyai, and A. Clardy. 2016. Economic Importance of Local Food Markets: Evidence from the Literature. *Journal of Food Distribution Research* 47(1): 57–63.

Food Marketing Institute. 2015. *U.S. Grocery Shopper Trends*, Food Marketing Institute: Arlington, VA. [www.fmi.org/research-resources/grocerytrends 2014].

Low, S., A. Adalja, E. Beaulieu, N. Key, S. Martinez, A. Melton, A. Perez, K. Ralston, H. Steward, S. Suttles, S. Vogel, and B. R. Jablonski. 2015. *Trends in U.S. Local and Regional Food Systems: A Report to Congress.* USDA, Economic Research Service. Administrative Report AP068. [http://www.ers.usda.gov/publications/ap-administrative-publication/ap-068.aspx].

MacDonald, J., J. Perry, M. Ahearn, D. Banker, W. Chambers, C. Dimitri, N. Key, K. Nelson, and L. Southard. 2004. *Contracts, Markets, and Prices: Organizing the Production and Use of Agricultural Commodities.* Agricultural Economic Report No. 837. U.S. Department of Agriculture, Economic Research Service.

MacDonald, J. M. 2015. "Trends in Agricultural Contracts" *Choices*, Quarter 3. [http://choicesmagazine.org/choices-magazine/theme-articles/current-issues-in-agricultural-contracts/trends-in-agricultural-contracts].

Martinez, S., M. Hand, M. Da Pra, S. Pollack, K. Ralston, T. Smith, S. Vogel, S. Clark, L. Lohr, S. Low, and C. Newman. 2016. *Local Food Systems; Concepts, Impacts, and Issues.* U.S. Department of Agriculture, Economic Research Service. Economic Research Report 97.

National Agricultural Statistics Service (NASS). 2012. *Farmers Marketing.* [http://www.agcensus.usda.gov/Publications/2012/Online_Resources/Highlights/Farmers_Marketing/Highlights_Farmers_Marketing.pdf].

USDA, Economic Research Service. 2016a. *U.S. Food Imports, Documentation.* [www.eers.usda.gov/data-products/us-food-imports.aspx].

_____. 2016b. *Agricultural Resource Management Survey.* [http://www.ers.usda.gov/data-products/arms-farm-financial-and-crop-production-practices/arms-data.aspx].

Vogel, S. and S. Low. 2015. The Size and Scope of Locally Marketed Food Production. *Amber Waves*. USDA Economic Research Service. [http://www.ers.usda.gov/amber-waves/2015-januaryfebruary].

Chapter 2

The Family Farm

In 2012 the United States had 2.1 million farms (USDA, NASS 2014). While slightly greater than the number counted by the USDA in 1992 and 1997 because of the addition of just over 300,000 small farms that were added to the Census's mailing list, it was a far cry from the nearly 7 million farms this country had in the mid-1930s (Figure 2.1). During the past eight decades, seven out of every ten U.S. farms disappeared. Acres of overall farmland have also declined—a measurement the Census of Agriculture refers to as land in farms. This category encompasses all land owned or rented by a farmer, whether cropland, pasture, grazing land, woodland, idled land, or land in conservation programs (see the Appendix for more details on the Census of Agriculture).

Since 1950 farmland acreage has declined 21% due in large part to the disappearance of "marginal" farms, the withdrawal of land from production, and the conversion of farmland to other uses (USDA, NASS 2014). The Census category that best represents land used to produce food is harvested cropland. This category includes land from which crops were harvested and hay was cut, land used to grow short-rotation woody crops including Christmas trees, and land in orchards, groves, vineyards, berries, nurseries, and greenhouses. Acres of harvested cropland have changed very little—only an 8% decline since the 1950s—and it has increased by more than 30 million acres since 1987. While many different types of farms exist, the production of agricultural goods has been, and continues to be, driven by a simple idea: increased yields, or the production of as much as possible from the same or smaller amounts of farmland. This shift, from many small farms producing small amounts of crops and livestock to fewer but larger farms producing copious amounts of crops and livestock, has been, and continues to be, the major trend in American agriculture (Hart 2003).

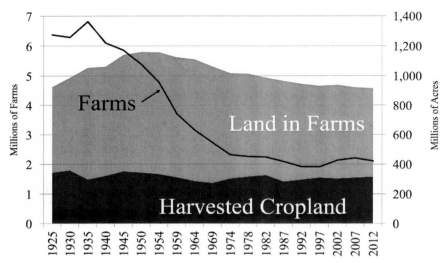

Figure 2.1   Number of farms, land in farms, and acres of harvested cropland in the United States since 1925. *Source*: Created by authors, data from USDA Census of Agriculture.

The USDA defines a farm as "any place from which $1,000 or more of agricultural products were produced and sold, or normally would have been sold, during the year" (USDA, NASS 2014). Accounting for inflation, and according to the U.S. Bureau of Labor Statistics, $1,000 of agricultural products in 1974 was equivalent to about $4,800 in 2012, which means that farms today that have met the $1,000 threshold have sold only about 20% as much as they would have needed to have sold to be considered a farm in 1974 (see the Appendix for farm definitions used over the years).

Historic settlement and land subdivision systems, along with agricultural practices and environmental characteristics, created distinctive patterns of increasing farm size westward from Ohio, through Illinois and Iowa, and into the Great Plains. As the settlement frontier expanded, west farm size increased as annual precipitation totals declined and forced farmers to acquire additional acres of cropland and rangeland to make up for lower yields and less forage for livestock. However, islands of larger-than-expected farm size occur in areas such as the Grand Prairie of east-central Illinois and western Indiana, where cash-grain corn and soybean farming originated after the Grand Prairie's wet soils were ditched and drained (Hart 1991). Within and west of the Rocky Mountains the pattern of gradual change gives way to cartographic chaos due in large part to climate and other biophysical conditions that change markedly over relatively short distances.

TYPES OF FARMS AND OWNERSHIP

Since the mid-1970s the average U.S. farm size has been relatively stable—though somewhat misleading—hovering around 430 acres. Over the past three decades the average number of cropland acres on farms has also changed very little, with the exception of a slight increase from 241 to 251 acres between 2007 and 2012 (USDA, NASS 2014). While the number of mid-sized crop farms has declined over the past few decades, the number of farms at both ends of the spectrum has grown, thus contributing to the relatively stable average (Figure 2.2).

In 1950, over 90% of all U.S. farms were 500 acres or smaller and were responsible for harvesting 70% of the nation's crops (U.S. Department of Commerce 1952). By 2012, 80% of the cropland was being harvested by the largest 20% of all farms. Further back in time and focusing on the state of Illinois, in 1925, 80% of the state's farms were 50 to 500 acres in size and harvested 93% of the state's cropland (U.S. Department of Commerce 1928). By 2012, 82% of the state's cropland was being harvested by only 28% of the state's largest farms, 14% of which were 500–1,000 acres, and 14.4% of which were over 1,000 acres in size (USDA, NASS 2014). The shift of

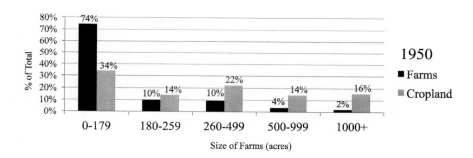

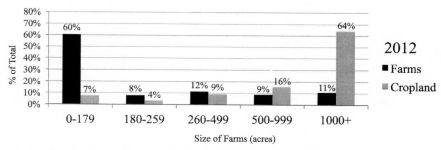

Figure 2.2 Farms by size and cropland acreage, 1950 and 2012. *Source*: Created by authors, data from USDA Census of Agriculture.

cropland ownership to the largest farms continues as high cropland prices generally prohibit all but the largest farmer or investor from purchasing land that comes up for sale (Hart and Lindberg 2014).

How has farm ownership been affected by these trends? This brings to question the general public's view of farm scale. Today's large farms produce much of the food that we consume. These farms use massive pieces of machinery, require large amounts of land, need large flows of cash, require access to costly inputs, and are increasingly owned by absentee owners— people who own the land but do not farm. To many people, big farming equates with corporate farming. This is only partially true.

The Census of Agriculture defines farms according to their legal status: Family or Individual, Partnerships, Family Corporations, Non-Family Corporations, and Other. Family or Individual farms are owned by one family. Partnerships are farms with multiple family farms where both parties have money invested in the business and they share management responsibilities as well as profits and losses. Family-corporate farms are no different than non-family corporate farms apart from requiring that principal stockholders be related by blood or marriage. "Other" farms include those managed as cooperatives, those that are a part of an estate or trust, or are part of an institution.

Despite the "industrious" or "corporate" appearance of even our country's most average-sized farms, family farms are still, far and away, the most common class of farm in America. In every agricultural census taken since 1978 the percentage of farms classified as "Family Farms" has been between 86% and 90% (USDA, NASS 2014). Farms classified as "Corporate" have been reported to comprise 2%–5% of the country's total number of farms, but even that statistic is misleading because many "corporate family farms" (88%–91% of all corporate farms) are housed in that category. Corporate family farms attained the physical and economic stature where "incorporating" the family's farm business for tax or other purposes was advantageous. Corporations also offer the benefit of protection of assets in the case of a lawsuit and can ease intergenerational fiscal transfers or gifts made to the next generation in the family.

The publicly perceived change in the number of family farms that has taken place since the mid-1990s has had little impact on farmland (including both cropland and ranchland). Family farms, partnerships, and corporate family farms comprise 97%–99% of all U.S. farms. With regard to land, 74%–81% of all U.S. farmland was owned and operated by family farms or corporate family farms in 2012. Including partnerships and other types of non-corporate farms increases the value to 91%–97% (USDA, NASS 2014).

In 1978 non-family corporate farms had the largest average acreage of the five farm types (920 acres) (U.S. Department of Commerce 1981). By 2012, corporate family farms had the largest average size (650 acres) and non-family corporate farm average size had shrunk to 465 acres. The average size of partnership farms more than doubled from 310 acres to nearly 640.

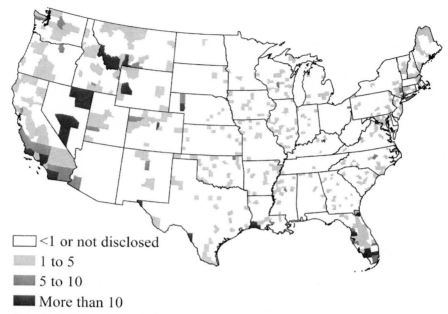

Figure 2.3 Percentage of farmland that is non-family, corporate-owned. *Source*: Created by authors, data from USDA Census of Agriculture 2012.

And while the majority of total cropland acres (77% in 1978 and 66% in 2012) were found on family or individual farms, the category that grew the most was family corporate farms—from 6.5% to 13% (USDA, NASS 2014). Many farms in this category once considered themselves individual or family farms, but as they grew in both physical size and profitability, the family business was incorporated (Raup 2002).

Non-family corporate farms tend to be found outside of the Great Plains-Corn Belt farming region (Figure 2.3). Clusters of non-family corporate farms are evident along the East and West Coasts, in the intermountain west, and in peninsular Florida, and are typically associated with the production of specialty crops such as fruits and vegetables. Most counties found in the greater than 10% category have relatively few farms of any type but enough non-family corporate farms to appear significant. As John Fraser Hart once said, "A large percentage of very little is still not very much" (Hart 2001, 533).

FARM PEOPLE

In 2012 women operated nearly 14% of all farms in the United States, up from only 5% in 1978 (U.S. Department of Commerce 1981). Women principal operators are found in all size classes of farms, although three-fourths of them operate farms with sales less than $10,000 per year (USDA, NASS 2014). If both principal and secondary operators are included, the number of women farm operators increases to nearly one million, or roughly half the farms in the nation. Compared with men, women farm operators on the average are older, are better educated, and are more likely to work off their farms at least on a part-time basis (Hoppe and Korb 2013).

Both men and women farm operators are getting older. The average age for men operators today is 57 years, for women, 59 years, and these averages continue to increase. Thirty-seven percent of farms in 1978 were operated by men younger than 45, but that share has fallen to only 18% today (USDA NASS 2014). The continued aging of the farm operator group reflects several trends, including better health care for farmers which allows them to control their own operations and live on their farms longer than once was possible (Hudson 2001).

The operation of farms continues to be largely under the direction of a white, male population. Ninety-five percent of the principal operators counted in the 2012 Census were white. Hispanic or Latino-origin farmers made up 3.1% of principal operators. African-American principal operators are found on just 1.6% of farms and Asian-Americans account for only 0.6%.

African-Americans made up nearly 15% of all farm operators in the United States as recently as 1900, but that number fell to 10% in 1950, and has continued to decline as the pressures on small farms continued to mount. Lack of access to credit and the uneven application of government assistance payments have been cited as adding to the difficulties that African-Americans have faced as farmers (Ficara 2006).

Hired farm workers make up one-third of all persons working on farms. Most of them are employed on the largest, most productive operations, and they are especially prevalent on dairy, fruit, and vegetable farms. Approximately one-half of all farm laborers and supervisors are Hispanic, while farm managers are typically non-Hispanic whites. Once a dominantly mobile labor force, most hired farm workers today live in the United States on a permanent basis and three out of five are U.S. citizens (USDA, ERS 2015).

OPERATING A CORN BELT FARM

The Midwest is the Corn Belt and is the agricultural powerhouse of the United States. The twelve states that comprise this region account for 40% of all U.S. farms, 37% of the nation's farmland, 57% of its total cropland, and 46% of the market value of all agricultural products sold. Corn and soybean acres in the Midwest account for 87% and 84%, respectively, of the nation's total. The Midwest has 78% of the country's hogs, 42% of its wheat acreage, 42% of its cattle, and 35% of its milk cows (USDA, NASS 2014).

One of the best ways to understand the farmers and farming in the Corn Belt is to examine the typical annual cycle of operations on one of its farms. Much of this cycle is determined by Mother Nature's grip loosening itself as winter turns to spring, and then the tightening as fall turns into winter. Other tasks, such as watching grain marketing options and keeping a watchful eye on what is happening with respect to crop health in other states and other countries, are year-round activities.

Illinois, which typifies the central Corn Belt, has 27 million acres of farmland and 88% of it is in crops (Smith 2014). Much the remainder is woodland and pasture in the southern and far western parts of the state. In 2012, Illinois's market value of agricultural products sold was $17.2 billion, $14.1 billion (82%) of which came from crops and $3.0 billion (18%) from the sale of livestock (USDA NASS 2014). Family farms dominate the agricultural landscape, with 99.5% of all farms in Illinois being family-owned, part of a partnership, or part of a family-owned corporation, and they account for 99.3% of the state's farmland. The average age of principal farm operators in the state was 57.8 years, nearly matching the national average of 58.3. Sixty percent of principal farm operators indicated that they worked "some" days off the farm at other jobs, with 37% working 200 days or more off the farm. Further, contemporary farmers often do not call farming their primary occupation; in 2012, 49.6% of all Illinois's farmers did not consider farming as their primary occupation.

THE ANNUAL CYCLE

January on the farm begins with budgetary concerns: visiting the local grain elevator on the first business day to pick up payment checks from grain delivered the previous harvest season, as well as paying off or making payments toward the farmer's operating loan (similar to a mortgage payment) at the local bank. Tax information (for the April 15th deadline) also begins to come in, and grain contracted for sale in January is hauled to the elevator from on-farm storage.

Seed orders for the upcoming year should also be finalized at this time, as should prepayments for the year's chemicals and fertilizer. Farmers with a winterized machine shed can finish fall harvest equipment cleanup, and they can also start putting equipment together or make repairs for the upcoming spring planting season. Meeting season has also started, a time when farmers attend local meetings that focus on marketing strategies, new tools and equipment, or new crop insurance products. Markets are watched on a daily basis, with each farm family having its own strategy on how to market grain they have in storage. At this time they must decide what they will be producing in the coming year taking into account local, national, and international supply and demand forces.

February continues the work that got under way in January, but the pace picks up. Tax documents continue to be collected and prepared, equipment repairs and grain hauling continue, and grain futures markets are watched more closely. February also sees federal crop insurance meetings convened by local USDA personnel. Toward the end of the month seed dealers begin delivering the year's orders of corn, soybeans, and wheat.

In March, dates such as the March 15th federal crop insurance deadline and the April 15th tax deadline are approaching. Federal crop insurance is a government program that began after the Dust Bowl, but remained largely experimental until it was reformed under the Federal Crop Insurance Act of 1980. The program protects farmers against losses that could occur during that crop year that are defined as unavoidable perils beyond the farmer's control. Nearly 63% of all cropland in 2012 was covered by federal crop insurance.

Markets continue to be watched and grain to fill March contracts is hauled to the elevator for sale. Of great importance at this time is the annual USDA prospective plantings report. Referred to by farmers as the "plantings intentions report," it indicates the expected amount of crop activity planned for the coming year, which often sets the baseline for prices farmers can expect to see. Needed repairs and other equipment-related issues should now be addressed if they have not been already. Seed deliveries are arriving for spring planting. Pre-plant application of anhydrous ammonia (nitrogen fertilizer) can begin any time the ground is warm enough. Burn-down spraying of weeds, pre-plant herbicide applications, and spring fertilizer applications will soon be made as well.

April is planting month, as soon as soil conditions allow. Final seed deliveries take place as farmers finalize their planting intentions. The application of pre-plant herbicides and fertilizers is in full swing and the scouting of planted fields is under way. Farmers look for weed pressures and insects that may necessitate further application of herbicides and/or insecticides. Post-application of herbicides can also begin once the crops have emerged.

From April into May, most farmers finish planting their corn crop and begin to plant soybeans after a pre-plant application of soybean herbicides is completed. Spraying of herbicides and insecticides on the corn crop after planting continues. With the year's row crops now in the ground, other activities begin to take place such as the mowing of ditches, equipment cleanup, and the scouting of fields for weed, insect, and disease problems.

Replanting may be necessary where spring rains drowned out what was already planted.

In June, post-emergent spraying work continues and spring planting equipment is cleaned and repaired before it is put away for the year. Some farmers begin readying their fall harvest equipment at this time as well. Ditch mowing and field scouting continue, as does grain hauling to empty any on-farm storage that will be needed come fall. For those farmers who planted winter wheat, its harvest begins in mid-to-late June, followed by the planting of "double-cropped" soybeans on the land from which the wheat was harvested. Double-cropping winter wheat and soybeans is a common practice where longer growing seasons allow farmers to harvest wheat in June and immediately plant a soybean crop that will be harvested in the fall.

Wheat harvest concludes in July, but the mowing of roadsides and ditches continues. Summertime is also when a "weather market" can develop—where farmers begin to see how the weather may impact their own crops and those of farmers around the county. Seasonal weather trends can create or destroy marketing opportunities, as well as influence projected prices for grain they have yet to harvest. Farmers also look ahead to pre paying for fall nitrogen and fertilizer (which would be applied after harvest for the subsequent year), as well as fuel costs for fall harvest needs. Fungicides are applied for both corn and soybeans as needed, and harvest equipment repairs and cleanup continue.

In August and September, the final preparations for fall harvest are under way, which include getting grain trucks ready for hauling this year's harvest, and preparing corn and soybean combines used to harvest the crop. Decisions for fall nitrogen and fertilizer applications are finalized, along with making pre payments on those. Just prior to harvest, farmers also make sure that storage bins are ready to receive the newly harvested grain.

When October arrives, harvest is in full swing. Aside from a day where harvest must be stopped because of rain (a perfect day to make needed repairs that have been put off because of the importance of harvest), there is not much else on a farmer's mind. But once fields have been harvested, fall fieldwork, fertilizer/lime/nitrogen/herbicide applications, and winter wheat planting begin. Seed dealers begin to visit farmers to get impressions of how their products fared, and to also take early seed orders for next year. As always, markets are watched closely, and harvest equipment is cleaned and stored, ready for winter.

November brings an end to the harvest season, and during this month the majority of fall fieldwork, along with nitrogen and fertilizer applications, comes to an end prior to the ground's freezing. Equipment is winterized and stored, and seed purchase decisions for next year are finalized and pre-paid.

Finally, December arrives. Grain markets continue to be watched for both the crop that was just harvest and next year's crop, which includes paying attention to what farmers' crops in South America look like during the southern hemisphere's summer, which can have huge implications for northern hemisphere farmers the following year with respect to what and how much of certain types of crops are planted. And because a farmer's work is never done, work will begin on equipment needed for next spring's fieldwork and planting season.

THE FUTURE

The annual cycle on a Corn Belt farm is not all that unlike other family farms in many other parts of the United States, deviating only based on what those farms produce. Family farms make up well over 90% of all farms with cropland, and in 2011 those farms accounted for 87% of the total value of U.S. crop production (MacDonald, Korb, and Hoppe 2013). But challenges and risks abound. It was estimated that with Corn Belt cropland selling for an average of $7,000 per acre in 2012, a farmer who has 600 acres of corn and 500 acres of soybeans would require $8 million of equipment, land, infrastructure, and other inputs to successfully operate a farm. For many, the debt and equity required to operate a farm of this average size is daunting and risky, with most or all of a family's monetary worth tied up in one large endeavor.

Even as agricultural production and land ownership have shifted to larger family farms, American farms and farmers continue to operate independent of large, corporate agro-businesses found in many South American and Asian countries. "They will continue to do so as long as they are able to limit and manage the financial risks associated with managing large and capital-intensive businesses, and as long as the strengths of family organizations—localized knowledge, quick and flexible adjustments to changed circumstances, and the incentives to act on each of those—remain necessary to crop production" (MacDonald, Korb, and Hoppe 2013, 50).

REFERENCES

Ficara, J. F. 2006. *Black Farmers in America*. Lexington: University Press of Kentucky.
Hart, J. F. 1991. *The Land That Feeds Us*. New York: W. W. Norton and Company.
Hart, J. F. 2001. Half a Century of Cropland Change. *The Geographical Review* 91(3): 525–543.

Hart, J. F. 2003. *The Changing Scale of American Agriculture*. Charlottesville: University of Virginia Press.

Hart, J. F., and M. B. Lindberg. 2014. Kilofarms in the Agricultural Heartland. *Geographical Review* 104(2): 139–152.

Hoppe, R. A. and P. Korb. 2013. *Characteristics of Women Farm Operators and Their Farms*. USDA Economic Research Service, Economic Information Bulletin No. 111.

Hudson, J. C. 2001. The Other America: Changes in Rural America During the 20th Century. In *North America: The Historical Geography of a Changing Continent*, edited by T. F. McIlwraith and E. K. Muller, Lanham, MD: Rowman and Littlefield.

MacDonald, J. M., P. Korb, and R. A. Hoppe. 2013. *Farm Size and the Organization of U.S. Crop Farming*. USDA. Economic Research Service, Economic Research Report No. 152.

Raup, P. M. 2002. Reinterpreting Structural Change in U.S. Agriculture. In *Economic Studies on Food, Agriculture, and the Environment*, edited by Canavari and others, Kluwer Academic/Plenum Publishers.

Smith Howard, K. 2014. The Midwestern Farm Landscape Since 1945, in *The Rural Midwest Since World War II*, ed. by J. L. Anderson, DeKalb, IL: Northern Illinois University Press.

U.S. Department of Agriculture, Economic Research Service (USDA, ERS). 2015. *Farm Labor Background*. [http://www.ers.usda.gov/topics/farm-economy/farm-labor/background.aspx#demographic].

U.S. Department of Agriculture, National Agricultural Statistical Service. 2014. *Census of Agriculture, 2012*. Washington, DC: USDA, NASS.

U.S. Department of Commerce, Bureau of the Census. 1928. *Census of Agriculture, 1925*. Washington, DC: U.S. Government Printing Office.

U.S. Department of Commerce, Bureau of the Census. 1952. *Census of Agriculture, 1950*. Washington, DC: U.S. Government Printing Office.

U.S. Department of Commerce, Bureau of the Census. 1981. *Census of Agriculture, 1978*. Washington, DC: U.S. Government Printing Office.

Chapter 3

The Corn Belt

The Corn Belt is America's best-known agricultural region. It is the undisputed home of the family farm and the stronghold of farming as a way of life. The Corn Belt is practically coextensive with the Middle West, an archetypal region in its own right that is often regarded as an exemplar of American values. Both regions are often called the Heartland, identifying not only their geographical position but also their strong influence in defining national character. No other region of the country comes close to playing such a role as this, nor does any other style of farming command anything like the respect accorded the "Corn Belt family farm."

The full-blown Middle Western ethos amounts to a good deal more than just the habit of raising corn, but the role this crop has played in the region's growth has been pervasive. As Cynthia Clampitt has written, "Whatever the future holds, it is fairly certain that corn will be at the center of the culture, economy, and foodways of the Midwest" (Clampitt 2015, 237). Why this should be so depends on a variety of historical and environmental influences that begin with the corn plant itself and with its adaptability as a feed grain for animals rather than just a food grain for people.

ZEA MAYS

Corn (*Zea mays, L.*)—known to most of the world as maize—is a human invention. Modern genetic studies have traced its origin to a single domestication event that took place in the highlands of southern Mexico about 9,000 years ago (Matsuoka and others 2002). Its wild progenitor is an annual grass called *teosinte* which is tall, leafy, and produces an abundance of what could be called "ears" on multiple stalks. The plant looks rather like corn, but the

23

ears do not. It is probable that deliberate human selection for ears producing multiple rows of kernels took place near the initial domestication site in southern Mexico. In a short span of time thereafter the plant was carried by human groups in nearly every direction—south to Chile, northwest to the arid lands of the Sonoran desert, east to the Caribbean islands, and eventually north into Canada. The many varieties of corn known today are a product of genetic isolation resulting from these long-distance dispersals into new territory. It is estimated that by the time of Columbus's arrival in 1492 there were 200 different cultivars of *zea mays* being grown somewhere in the Americas (Beadle 1980).

Of the many variations on corn, such as ears with blue kernels or red kernels, popcorn types, sweet corn, and drought-adapted varieties, two lines stand out as the foundation of what could be called Corn Belt corn. One is the variety often called Indian corn which has eight rows of kernels on a long, slender, tapered ear. Also known as New England Flint corn, it descended from maize that was introduced into the arid southwestern United States from Mexico. From there the crop spread eastward through the agencies of human migration and trade to become established as a food crop among native peoples in the Mississippi River Valley about 1,400 years ago. Eventually it reached New England.

Other types of corn produce what have been called hand-grenade-shaped ears. This second group of maize varieties sprang from the same initial domestication event in the southern Mexican highlands, but the paths of dispersal were directed more to the south and east. Varieties known to have been grown early in the Andes, the Amazon lowlands, and on the Caribbean islands share the hand-grenade-shaped ear trait. Corn of this type was described in an eighteenth-century account of agriculture in Virginia (Beverly 1772). The ears were short, thick, and had 18–22 rows of dented kernels which were soft and easily chewed by cattle and hogs. Known as Dent corn, it was in production as a feed grain for animals in the eastern states by 1800.

The importance of these two types of corn is not so much their differences as their ability to cross freely with one another. Although the initial crosses probably were by chance, resulting from two kinds of corn growing in the same field, it was soon learned that a superior product resulted from deliberately growing the two together. This was the origin of open-pollinated Corn Belt Dent corn, the variety that sustained the growth and expansion of the Corn Belt through the nineteenth century until well into the twentieth. Corn Belt Dent corn had many rows like the Southern Dent and it had a long cob like the New England Flint. It was a bountiful crop, especially when it was introduced into new lands where corn had not been grown before. Sacks of Corn Belt Dent were part of the baggage carried by pioneer settlers who crossed the Appalachians and settled in the Ohio Valley during the late eighteenth century (Anderson and Brown 1950).

EMERGENCE OF THE CORN BELT

The first large corn crops were produced in "five islands" of good agricultural land that lay west of the Appalachians: the Nashville Basin of Tennessee, the Pennyroyal Plateau along the Tennessee-Kentucky border, the Bluegrass of Kentucky, and the Virginia Military District and the Miami Valley in southwestern Ohio (Hudson 1994). The pioneer settlers of all five of these areas had moved west from Pennsylvania and Virginia in the latter decades of the eighteenth century via well-known routes such as down the Great Valley and through Cumberland Gap. They brought the practices of corn-livestock agriculture with them, including the ears of Flint and Dent corn that would be the foundation of their crops from that time forward.

The corn crops raised in those five islands of good land were fed to livestock which, once fattened, were driven back across the mountains to markets in Baltimore, Philadelphia, and New York (Henlein 1959). Hogs were driven east as well, but most of them became cured meat, pork lard, and hides that were shipped in barrels downriver on the Ohio and the Mississippi to eventually reach markets on the East Coast or overseas. By the 1820s corn-planting farmers were spreading west, up the river valleys in Indiana and Illinois,

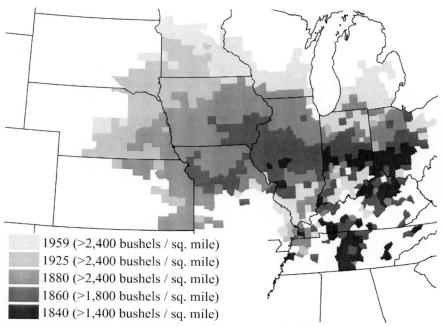

1959 (>2,400 bushels / sq. mile)
1925 (>2,400 bushels / sq. mile)
1880 (>2,400 bushels / sq. mile)
1860 (>1,800 bushels / sq. mile)
1840 (>1,400 bushels / sq. mile)

Figure 3.1 Historical expansion of the Corn Belt, 1840–1959. *Source:* Created by authors, data from USDA Census of Agriculture.

following the proven strategy of raising a corn crop for the purpose of fattening cattle and hogs for market.

In 1840, when the first Census to include agricultural statistics was taken,
the lower Wabash Valley and the Sangamon country near Springfield,
Illinois, had been added to the list of major corn producers (Figure 3.1). Until
the 1850s, when railroad construction began in earnest, rivers remained as
the only practical route to market. Tracts of fertile bottomland along rivers
like the Wabash, Missouri, and Illinois attracted new would-be corn farmers
who moved there from the Ohio Valley, lured by stories of the large yields
these new lands in the West could produce. With the appearance of railroads,
agricultural settlement spread more rapidly across the countryside, no longer
confined to the zone within access of a navigable river. Most counties of central Indiana, Illinois, and southeastern Iowa had been added to the Corn Belt
by 1860. By that time large corn crops had spread up the Missouri Valley to
the eastern edge of Kansas as well.

The accessibility of fertile lands and improvements in market access
explain part of corn's rapid expansion to the west, but the crop's history and
genetic makeup played a role as well. Corn is a C4 grass, which refers to a
type of photosynthesis in which the plant makes maximum growth during the
hot summer months. Corn is an efficient water user that can withstand summer dry periods of at least moderate duration. As an annual grass, corn grows
rapidly in stature when the long daylight hours of early summer occur. Corn
can adapt to longer growing-season-day-lengths, such as would be encountered if seed from more southern latitude is planted farther to the north, but
it often takes several generations of reproduction to make such a northward
shift. For all of these reasons it was much easier for corn farming to spread
straight west from the Ohio Valley than it was for it to spread in any other
direction. This was to be the Corn Belt's general direction of expansion until
well into the twentieth century.

Corn crops continued spreading with the western frontier across Nebraska
and Kansas, but the plant's dryland limit was encountered when a series of
droughts stopped its westward advance in the 1890s. Kansas farmers who had
raised corn abandoned the crop and either moved back east or switched to
wheat farming, which was much less risky in a drought-prone environment.
It was not until irrigation was introduced on a large scale in the 1950s that
corn production found a home in the Great Plains.

Once farmers became dedicated to growing a corn crop, the natural desire
to increase yields and improve the grain became a focus of experimentation
and competition (Mosher 1962). Although corn yields had been surprisingly
large in the early years of crop agriculture in the Middle West, those early
yields were not sustained. In longer-established areas of corn culture the

yields actually declined over time. Farmers applied animal manure as fertilizer and they saved corn kernels from the ears of plants that had been most
productive for planting the next year, but few yield increases repaid these
efforts.

The corn crop of the United States grew substantially during this period
and it was entirely due to the Corn Belt's expanding area, with more farmers
planting more acres to corn every year. By 1880 the Corn Belt had spread
northward across Iowa and by 1925 a new band of counties across Nebraska,
South Dakota, and Minnesota had been added. Corn production in the United
States increased by 40% from 1880 to 1925, but the number of acres used
to grow the crop increased by an almost identical 39%. Yields fluctuated
between years of good weather and years of bad, but they showed no consistent upward trend despite the efforts to produce more corn per acre.

What the farmers did not understand was that the corn varieties they were
planting were a mixture of all of the traits that had come down from the early
Flints and Dents. Every corn plant in every field differed only by chance from
the one growing next to it. Around 1900 the rediscovery of genetic principles
first proposed by Gregor Mendel (1822–1884) were applied to corn breeding,
and with spectacular results. Guaranteeing that the pollen falling on the silk
of an ear of corn came only from the tassel of that same plant, and repeating
this "selfing" over a number of generations, resulted in backcrossing to original ancestors. The inbred ears produced little grain but they represented pure
lines that could then be crossed. Subsequent crossing of two of the inbred
lines captured the hybrid vigor inherent in the long history of geographic
isolation that had produced the divergence between New England Flint and
Southern Dent types.

By 1940 hybrid corn, as the new crop became known, had become the standard of the Corn Belt (Griliches 1957). Yields increased steadily from that
time forward and corn production began to expand northward with the guidance of new hybrid corn varieties more suited to a shorter growing season
with longer days. Farmers in southern Michigan, Wisconsin, and Minnesota
were raising large crops of corn for grain by the 1950s. The overwhelmingly
most important use made of the crop was for livestock feed on the farms
where the crop was raised.

Fields were planted at a rate of about 12,000 plants per acre which allowed
the rows to be far enough apart to permit cultivation in both directions. The
crop was planted in May, harvested in October, and stored on the farm as
ear corn for later use. The "corn picker" was a tractor-mounted harvesting
machine that plucked the ears off the stalks and kept the ear and its tightly
held kernels intact for storage (Figure 3.2). Yields of 80–100 bushels per acre
were considered good up until the early 1970s.

Figure 3.2   Tractor-mounted corn picker harvesting ears of corn in the 1970s. *Source*: Photograph courtesy of author.

SOYBEANS

Although corn remained the unchallenged focus of Midwestern crop production through the first half of the twentieth century, the elimination of horses as draft animals freed some farm acreage for new uses. The time was ripe to introduce another cash crop. Soybeans were tried experimentally on American farms in the 1930s, at about the same time that hybrid corn was first being adopted. The two crops developed an association that has endured ever since.

Soybeans (*Glycine max L.*) are native to northeastern China where they were domesticated about 9,000 years ago. References to this crop, which produces round, white "beans" in a greenish pod, appear in written accounts from China and Japan, dating to the middle of the first millennium BCE. Food items made from soybeans, such as tofu, natto, and soy sauce, were being produced in Japan by the early 1600s (Shurtleff and Aoyagi 2014). But despite the plant's long history of use in Asia, it was barely known in North America until the end of the nineteenth century (Piper and Morse 1923).

The USDA imported several soybean varieties from Japan at that time and the seeds were distributed to agricultural experiment stations in the South and the Middle West. One of the special attractions the soybean had was its ability to fix atmospheric nitrogen in the soil. Bacteria growing in nodules on the roots of the soybean plant allow the plant to provide some of its own nutrients. Soybeans became known as a good crop to grow in soils that were a little too poor for obtaining a large corn yield. The nitrogen supply also could carry over to the next year's growing season, which made it advantageous to plant corn on land that had grown soybeans the year before. Corn is a prodigious nitrogen feeder. The two crops have been grown in association on farms of the Corn Belt ever since.

Soybeans thrived in the Illinois's Grand Prairie, the Des Moines Lobe in Iowa, and the Maumee Plain of northeast Indiana and northwest Ohio. Soybeans comprised no less than one-third of all harvested cropland in those areas by 1960 and from those early centers of production spread throughout today's Corn Belt (Figure 3.3).

The major use of corn was in its role as a feed crop on the farm where it was grown. Soybean forage was useful on the farm as a hay for livestock, but the beans had no immediate use. Rather, they were a valuable cash crop if trucked to a crushing plant where oil and meal were extracted from the beans. Soybean oil is a major food ingredient, especially in salad dressings and other

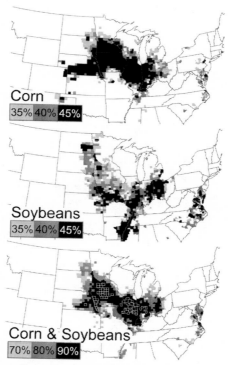

Figure 3.3   Corn, soybeans, and combined acreage as a percentage of harvested cropland in 2012. *Source:* Created by authors, data from USDA Census of Agriculture 2012.

manufactured food products, while soybean meal is consumed primarily as poultry feed. Demand for both increased dramatically after the 1930s. Soybeans were being raised on 4.3 million acres of U.S. farmland in 1939, a land use which increased to 22.1 million acres in 1959. Outside the Corn Belt lands in the alluvial Mississippi Valley and portions of the Southeastern Gulf Coastal Plain support millions of bushels of soybeans raised for poultry feed.

EXPANSION OF THE REGION

Corn began to shift westward once again in the mid-twentieth century (Figure 3.4). The most commonly recognized western edge of corn production had been the 100th meridian, which runs down the middle of the Dakotas and Nebraska, roughly following the 20″ rainfall line. West of this line it had been possible only to raise dryland corn that was cut and fed to feedlot cattle. Ditch- or gravity-fed irrigation had begun near Garden City, Kansas, in 1881, but never became widespread because it requires a nearby stream with a steady supply of water. Modern center-pivot sprinkler irrigation became feasible when powerful water pumps were invented that could draw water up

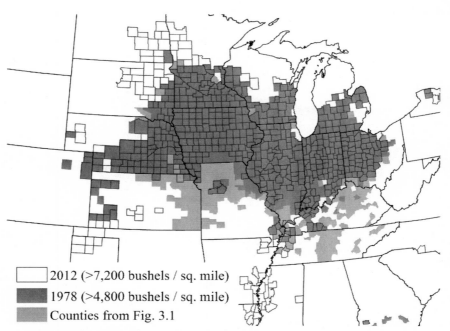

2012 (>7,200 bushels / sq. mile)
1978 (>4,800 bushels / sq. mile)
Counties from Fig. 3.1

Figure 3.4   Recent expansion of the Corn Belt, 1978–2012. *Source*: Created by authors, data from USDA Census of Agriculture.

hundreds of feet from the underlying High Plains/Ogallala Aquifer. Irrigation atop the aquifer soon became widespread throughout Nebraska, western Kansas, eastern Colorado, and in both the Oklahoma and Texas panhandles (Bremer, 1976; Green, 1973; Sherow, 1990).

Between 1950 and 2009 Kansas farmers increased their corn production from 85 million bushels to 561 million bushels largely because of the increased adoption of irrigation technology. The growth slowed after 1980, however, because of higher energy costs and decreases in water availability. Some farmers, beginning in the mid-1980s, enrolled a portion of their lands in the federal Conservation Reserve Program which returned erodible lands to grass covers (see Chapter 10). In 1950, counties west of the 100th meridian and considered a part of the contemporary Corn Belt planted 48,000 acres of irrigated corn that produced 17 million bushels. Today those same counties grow 2.6 million acres of irrigated corn that yield just less than 600 million bushels.

The development of new hybrid corn varieties has also expanded the Corn Belt's northwestern fringe (Napton and Graesser 2012). In east-central North Dakota corn must mature in 80–90 days to survive the shorter growing season, whereas corn grown in central Illinois has over 120 days in which to mature. The short-season hybrids also have a lower yield per acre. Both of these limitations may prohibit the future northward expansion of large, high-yielding corn acreages.

THE SHIFT FROM CORN-LIVESTOCK TO CASH GRAIN

The most important change in Corn Belt agriculture during the past 75 years has been a reorganization of the farming system itself (Hart 1986, 2003). Corn acreage in Minnesota and Iowa has increased only slightly, from just over 40% of the region's harvested cropland in the early part of the twentieth century to about 50% in recent years. Soybeans have replaced crops such as wheat, oats, and hay, which once were found on over half of the cropped land. As farmers stopped raising livestock, a greater proportion of their annual sales came from the sale of their crops. And as livestock disappeared from Corn Belt farms, the need for pastureland and hayland also declined, creating even more acres for corn and soybean production (Napton 2007).

In 1900 cattle were found on virtually all Indiana, Illinois, and Iowa farms, and hogs were found on three-fourths of them. But by 2012, the percentage of Corn Belt farms with cattle or hogs dropped to 27% and 5%, respectively. By and large, Corn Belt farmers do not want the extra work of raising livestock and instead focus their time and money on growing corn and soybeans. But even with this massive decline in the number of cattle and hog farmers in the region, the number of hogs has increased from 37 to 45 million head and cattle have increased from 29 million to 38 million (Chapter 6).

NEW MARKETS

The United States is the world's largest exporter of corn. Corn grain exports represent a significant share of total U.S. corn production and make a large contribution to the nation's agricultural trade balance for agricultural commodities. U.S. corn exports increased from 13 million metric tons in the early 1970s to a record 62 million by 1980 largely because of strong demand in Russia, Japan, and Europe. Exports have fluctuated since that time and other countries, especially Brazil, have become major corn exporters as well. The diversion of a large portion of the U.S. corn crop into ethanol production limits the amount that can be exported.

Corn and soybeans, while grown in rotation with one another, have widely different end uses, none of which are necessarily of any great concern to the farmers who grow them (Figure 3.5). Today's cash-grain Corn Belt farmers are interested in maximizing their yields and moving their crop from the field to the purchaser as quickly as possible once harvest begins. Chances are that unless farmers deliver their corn or soybeans to a facility that manufactures a specific product, such as an ethanol or biodiesel, they likely will never know what became of their year's agricultural bounty.

Soybeans are the world's most important source for protein in animal feed and the second largest of type of vegetable oil. Nearly half of the soybeans produced in the United States are exported, either as whole beans or soybean meal. China, Mexico, Japan, Taiwan, and the EU countries are major customers. The U.S. poultry industry is the top consumer of both corn and soybean meal used for the feeding of livestock, although beef and pork production together claims a larger share of the total corn crop that goes for feed.

Corn and soybeans are the main feedstocks used to manufacture the biofuels now produced in the United States. Corn ethanol is blended with gasoline to produce E10 (10% ethanol and 90% gasoline) and soybeans are used to make biodiesel (methyl ester). Although only about 2% of soybeans are used to produce soy diesel, some 38% of the corn crop goes into ethanol. One bushel of corn produces about 2.75 gallons of ethanol which, at today's average corn yield, translates to more than 500 gallons of ethanol per acre. In 2014 the United States produced more than 14 billion gallons of ethanol, thus requiring about 28 million acres of land, roughly equal to the entire cropland acreage of Iowa. Byproducts of ethanol production include distillers grains, which are a valuable source of livestock feed; they, too, are exported.

Corn-based ethanol production surged during the latter part of the 1990s and early 2000s, but has leveled off since 2008. Biofuel production is

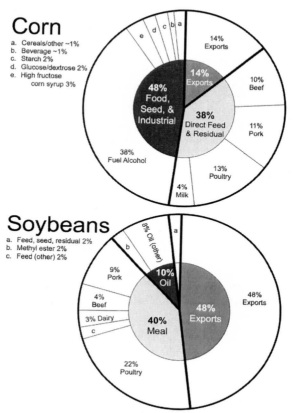

Figure 3.5 Uses for corn and soybeans. *Source*: Created by authors, data from USDA Census of Agriculture 2012.

governed by the U.S. Renewable Fuel Standard that was created as part of the Energy Policy Act of 2005 and further augmented by the Energy Independence and Security Act (EISA) of 2007 (EPA 2015). The program had a goal of producing 36 billion gallons of renewable fuel by 2022. Corn-based ethanol, which is considered a conventional biofuel, is projected to follow EISA guidelines to reach a plateau of 15 billion gallons with the rest to be provided by cellulosic ethanol and advanced biofuel production.

CHANGING TECHNOLOGY

Since 1925 harvested cropland acreage in the twelve Midwestern states has increased by only 4% and has actually declined during the past decade as farmers focus their efforts on getting more yield from their best lands. Most farmers have embraced the concept of genetically modified organism (GMO) seeds since their commercial introduction in the late 1990s (USDA 2015). Genetically engineered crops have been created to be herbicide-tolerant (HT), insect-tolerant (Bt), or both—a process known as "stacking" traits. HT crops are able to survive exposure to chemicals that would otherwise kill them along with the weeds the farmers want to eradicate. Bt crops contain a soil bacterium gene, *Bacillus thuringiensis,* which releases a protein toxin for killing insects. Stacked seed varieties now make up 92% of the corn and 94% of the soybeans grown in the United States (USDA 2015).

One of the more common HT varieties is Roundup Ready® Soybeans produced by Monsanto. "Roundup ready" means that the seeds have been engineered to be resistant to glyphosate, also manufactured and sold by Monsanto, which is the active chemical that kills weeds and grasses. Because the soybeans have been genetically altered to withstand glyphosate application, the weeds are killed but the soybeans are not. Herbicide applications make it unnecessary for farmers to make multiple trips through their fields cultivating weeds during the early growing season. This not only saves on time, fuel costs, and emissions, but fewer trips with heavy machinery also reduce soil compaction which increases yields and helps control soil erosion. Better seed, as well as the availability of herbicides, insecticides, and fungicides, has helped triple corn yields over the past half-century.

Along with advances in crop hybridization, genetics, and other biotic and abiotic inputs, the increased size and sophistication of farm implements (Figure 3.6) has helped farmers increase yields and improve the conditions of their land (Anderson 2009). Global positioning system (GPS) receivers mounted on top of tractors, along with other spatial technologies, are part of what is termed "precision agriculture." GPS signals steer tractors and combine harvesters in a precise course across a field which reduces fuel consumption by avoiding unnecessary overlap caused by human error. GPS technologies are also used to correlate soil or plant survey data with application rates of chemicals and fertilizers, a system known as variable rate application. For example, farmers sometimes perform a post-harvest soil survey, sampling across a field in order to map varying soil acidity or nutrient levels. The map then becomes an electronic guide automatically regulating the application rate of nutrients to the field.

Precision applications of fertilizer and soil nutrients reduce the amount of chemicals applied to the field, saving money and helping to reach conservation goals (Auch and Laingen 2015). Tractors, combines, and large grain wagons have also begun utilizing rubber track systems instead of wheels. With machinery's increase in size and tillage less common, soil compaction, which could lead to yield decline, has become a concern. Tracks allow the weight of increasingly large machinery to be spread out over a larger surface area, thereby reducing compaction.

Corn Belt farmers have also adopted soil and nutrient conservation tillage techniques known as no-till and strip-till. Instead of plowing up a field

Figure 3.6 A typical corn harvest; an 8-row combine unloading grain into a grain cart. Grain will then be unloaded into a waiting semi-tractor/trailer (background) which will take the grain to a local elevator. Photograph courtesy of author.

after harvest is complete to ready it for planting the following spring, farmers now leave crop stubble and other types of crop residue undisturbed on the soil surface. This not only lowers the potential for soil erosion caused by wind and water, but also acts like a gardener applying mulch to a flower bed as it increases both soil moisture and temperature. Added soil moisture benefits the next spring's crop and higher temperatures promote faster seed germination.

Leaving crop residues in place also means that more nutrients are left in the field and that the soil's organic-matter content will be increased. Individual plants themselves are now planted more densely than in the past when plants were spaced far enough apart to allow cultivation both directions. Thirty-inch row spacing and 5″ to 6″ spacing of plants is now the norm. Where farmers in the 1950s were planting 12,000 seeds per acre, farmers today may plant as many as 40,000 seeds per acre.

Today's Corn Belt farmers are part of a globalized agricultural market. They need to be keenly aware of market demands from countries such as China, which has quickly become the U.S. number-one importer of soybeans. They need to know about the weather affecting crop yields in countries like Argentina and Brazil which, together, produce and export more soybeans than does the United States. Technological innovations in drainage, irrigation, crop genetics, fertilizer, chemical applications, GPS, and farm implements as well as changes in government farm policies have created a landscape of far greater complexity, but also one that is increasingly recognized as an endowment of nature that must be sustainably managed.

REFERENCES

Anderson, E., and W. L. Brown. 1950. The History of Common Maize Varieties in the United States Corn Belt. *Journal of the New York Botanical Garden* 51: 242–267.

Anderson, J. L. 2009. *Industrializing the Corn Belt: Agriculture, Technology, and Environment, 1945-1972.* DeKalb, IL: Northern Illinois University Press.

Auch, R. F., and C. R. Laingen. 2015. Having it Both Ways? Land Use Change in a U.S. Midwestern Agricultural Ecoregion. *The Professional Geographer* 67(1): 84–97.

Beadle, G. W. 1980. The Ancestry of Corn. *Scientific American* 242(1): 112–19.

Beverly, R. 1855. *The History of Virginia [1772].* Richmond: J. W. Randolph.

Bremer, R. 1976. *Agricultural Change in an Urban Age: The Loup Country of Nebraska, 1910-1970.* Lincoln: University of Nebraska Press.

Clampitt, C. 2015. *Midwest Maize: How Corn Shaped the U.S. Heartland.* Campaign: University of Illinois Press.

Environmental Protection Agency. 2015. Program Overview for Renewable Fuel Standard Program. [http://www2.epa.gov/renewable-fuel-standard-program].

Green, D. E. 1973. *Land of the Underground Rain: Irrigation on the Texas High Plains, 1910-1970.* Austin: University of Texas Press.

Griliches, Z. 1957. Hybrid Corn: An Exploration in the Economics of Technological Change. *Econometrica* 25: 501–523.

Hart, J. F. 1986. Change in the Corn Belt. *Geographical Review* 76(1): 51–72.

_____. 2003. *The Changing Scale of American Agriculture.* Charlottesville: University of Virginia Press.

Henlein, P. C. 1959. *Cattle Kingdom in the Ohio Valley.* Lexington: University of Kentucky Press.

Hudson, J. C. 1994. *Making the Corn Belt.* Bloomington: Indiana University Press.

Matsuoka, Y., Y. Vigoroux, M. Goodman, J. Sanchez, E. Buckler, J. Doebley. 2002. A Single Domestication for Maize Shown by Multilocus Microsatellite Genotyping. *Proceedings of the National Academy of Sciences* 99(90): 6080–6084.

Mosher, M. L. 1962. *Early Iowa Corn Yield Tests and Related Later Programs.* Ames: Iowa State University Press.

Napton, D. E. 2007. Agriculture. In *The American Midwest: An Interpretive Encyclopedia,* edited by R. Sisson, C. Zacher, and A. Cayton. Bloomington: Indiana University Press.

Napton, D. E., and J. B. Graesser. 2012. Agricultural Land Change in the Northwestern Corn Belt: 1972-2007. *Geo-Carpathica Anul XI*(11): 65–81.

Piper, C. V., and W. J. Morse. 1923. *The Soybean.* New York: McGraw-Hill.

Sherow, J. E. 1990. *Watering the Valley. Development along the High Plains Arkansas River, 1870-1950.* Lawrence: University Press of Kansas.

Shurtleff, W., and A. Aoyagi. 2014. *Early History of Soy Worldwide to 1899.* Lafayette, CA: Soyinfo Center.

USDA. 2015. Adoption of Genetically Engineered Crops in the U.S. [http://www.ers.usda.gov/data-products/adoption-of-genetically-engineered-crops-in-the-us.aspx].

Chapter 4

Wheat and Grains

Wheat emerged as a staple of the human diet nearly 10,000 years ago following its domestication in the highlands of the Eastern Mediterranean. The domestication of plants allowed people to abandon their highly mobile lifestyles as hunter-gatherers to follow a more settled, sedentary life. The tending of crops demands some level of attachment to a place and also invites experimentation to improve crop yields. Societies changed fairly rapidly once these habits were followed and about 7,000 years ago the great civilizations of the Middle East and Mediterranean were rising, one after another.

Wheat is one of the first crop plants that has been identified as the product of human care and attention. *Triticum*, a genus of the grass (*Poaceae*) family, offered several useful plants for selection. Einkorn (*T. monococcum*), Emmer (*T. dicoccon*), Spelt (*T. spelta*), and Common Wheat (*T. aestivum*) emerged in the highlands of Syria, Turkey, and Iraq approximately 8,000 years ago. The genetic variety present in that region remains important today because it allows the development of new strains of wheat from a diverse genetic stock.

Other genera of the grass family include *Hordeum*, which is notable for the species *H. vulgare*, which has the common name barley; and *Secale*, which includes *S. strictum*, which is cultivated rye. Added to this variety of important crop plants is *T. durum*, an especially hard-kerneled variety of wheat that is most often grown for pasta. Durum was developed from strains of Emmer grown about 7,000 years ago (Weiss and Zohary 2011).

The first breads made from these wheats probably were of the unleavened variety. Yeasts, which are fungi that are present in the air or as a natural contaminant in wheat flour, can greatly improve the flavor and texture of bread when they are added. The common use of yeast is at least 4,000 years old, even though the biology of microbial fermentation was not understood until Louis Pasteur described the process in the 1860s. Yeasts and the process of fermentation also became used with barley which led to the production of beer. Bread wheats, pasta wheats, and grains for yeast fermentation used in making whiskey, beer, and wine thus emerged from a single group of crop plants some thousands of years ago.

Wheat and the other small grains were taken from the Middle East by traders and immigrants who traveled east to India, north to the steppes of Russia, and west to the Mediterranean. Wheat was established in Europe more than 6,000 years ago and in China (where rice already had been domesticated) more than 4,000 years ago. Wheat became an essential bread grain in all of these areas and, of course, remains so today.

Wheat and its close relatives follow the so-called C3 pathway for fixing carbon in the plant tissue as part of the process of photosynthesis. C3 plants grow rapidly in the cool months of the year and then more slowly in the warmer, high-sun months when the grain matures. Wheat is thereby called a "cool season crop," but that does not mean it can be grown only in cool climates. Wheat genetics slowly changed over thousands of years in Europe as the crop moved northward, into latitudes of longer day lengths but shorter growing season.

Two distinct approaches to raising wheat eventually came about. One was to plant the crop in the fall, allowing it to grow in the cool, low-sun months, letting it lie dormant during winter, and then to harvest it early the next summer. Wheat grown in this manner is known as winter wheat and it is the common type of wheat culture found where winters are mild and summers are hot. In northern latitudes, where summer days are long but the growing season is short, the strategy is to plant the crop as early as possible in the spring and harvest it in the fall; this is termed spring wheat culture.

THE EMERGENCE OF AMERICAN WHEAT REGIONS

Because wheat production and consumption were so widespread in Europe, it is not surprising that the crop was brought to the Americas by all of the colonizing nations. Wheat grew well under a variety of conditions present in the Western Hemisphere, but it was new to the native peoples, most of whom relied on maize (corn) for their bread grain. The Spanish quickly adopted the New World maize-based agriculture, but the English and French retained their preference for wheat even though maize was grown in abundance in the areas they colonized.

By the mid-1700s wheat fields dotted the better lands of the eastern seaboard colonies in North America, but wheat production per acre was low. It took the efforts of thousands of farmers and laborers to produce somewhat less than one million bushels of wheat per year even into the early nineteenth

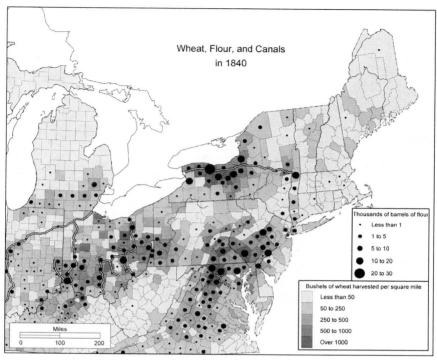

Figure 4.1 Wheat, flour, and canals in 1840. *Source*: Created by authors, data from USDA Census of Agriculture.

century. As population increased and people pushed westward to new lands, the sowing of a wheat crop was often one of the first tasks to be accomplished when settlers arrived in a new place.

Statistics on crop production were not recorded in the U.S. Census until 1840, well after the colonial period had ended but before the main thrust of Euro-American settlement into the Mississippi Valley had begun. The contrasts between good land and poorer land are evident by that date (Figure 4.1). Smooth, rolling lands in New York, Pennsylvania, Maryland, and Virginia were the "breadbasket of America" in the early decades of the nineteenth century. This was some of the best agricultural land in the United States and it lay close to the cities of the East Coast. New England, with its large population but poorer soils, already had become a net importer of wheat from the better lands to the south.

Trade in wheat between regions was more likely trade in flour rather than in the raw, unmanufactured grain. Grist mills—little more than a pair of huge, polished grinding stones slowly turning on the power of a water wheel—were small-scale industries built along streams both large and small in the areas where wheat was grown. Getting the flour to market was difficult in this era when only river and canal boats were available to handle bulk transportation. It is thus not surprising that the first westward advances of wheat raising and flour milling are closely associated with the building of canals.

The largest concentration of wheat fields and flour mills in 1840 was in the valley of the Genesee River around the city of Rochester, New York. Five counties near Rochester produced nearly one million barrels of flour from the local wheat crop in 1840. The Erie Canal, completed in 1825, provided a means to carry the flour to New York City, where the product was consumed or resold for export elsewhere. The importance of both good land and transportation access was revealed in the Genesee Valley's early success in wheat and flour production (McKelvey 1949).

A third wheat-producing region was emerging in the Ohio Valley in 1840. Transportation access was afforded by the Ohio River and Lake Erie, but a wave of canal-building efforts, following on the success of the Erie Canal, extended water transportation inland in both Ohio and Indiana. The Ohio and Erie Canal connected flour mills in eastern Ohio with Lake Erie and the Ohio River, and the Miami and Erie Canal did the same in western Ohio. The Wabash and Erie Canal extended wheat production far to the north in Indiana, and the Whitewater Canal made southeastern Indiana accessible to the great river port of Cincinnati.

Canals turned out to be only a temporary expedient. A better alternative came in the 1850s when railroad lines were built inland from the Great Lakes ports as the wheat frontier moved west to Michigan and Wisconsin and into the Mississippi Valley (Figure 4.2). New concentrations of flour milling based on local wheat production emerged around St. Louis and in southern Michigan. The dispersed nature of the flour mills matched that of wheat production, showing that the milling industry was still a local enterprise. That pattern changed when Minneapolis became the nation's largest flour-milling city. Grain millers and railroad builders stimulated the development of the Northwest as a massive wheat-producing region tributary to Minneapolis.

Hard, red spring wheat was the main type raised on the farms that fed grain to the flour mills of Minneapolis. While it was high in protein content, the flour ground from it typically was dark in color and was flecked with impurities. In the old system of flour milling that had prevailed since the early nineteenth century, wheat was ground only once. By the 1870s flour millers in Europe had perfected new methods that employed several reductions of the grain from wheat to flour using steel rollers. Sifting methods were employed to produce a fine, white-colored flour. Millers in Minneapolis, including the Pillsbury company and the Washburn-Crosby concern (Gold Medal Flour), invested in the new process roller mills. By the 1880s Minneapolis millers

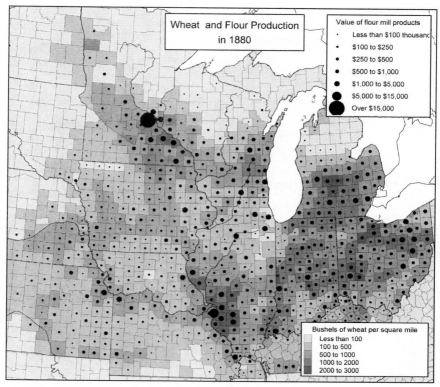

Figure 4.2 Wheat and flour production in 1880. *Source*: Created by authors, data from USDA Census of Agriculture.

had captured a large share of the national bread flour market and the city ruled for years thereafter as the nation's leading flour-milling center (Kuhlman 1929).

All of the varieties of wheat grown in the United States and Canada were brought from Europe at various times. In the late 1870s German Mennonite farmers who had been living in South Russia brought to central Kansas the grains of several wheat varieties they had been planting. It was most likely the Mennonite immigrants who introduced the varieties of hard, red winter wheat to Kansas and thereby supplied a key ingredient for yet another large expansion of production. By 1900 the winter wheat zone had grown in its production intensity and increased in area until it stretched from southern Nebraska to central Oklahoma. Winter wheat became the dominant crop of the southern Great Plains by 1920. Kansas farmers had experimented for years with wheat, corn, and livestock, but the new wheat varieties ended the uncertainty over what would be the specialty of the state's farmers (Malin 1944).

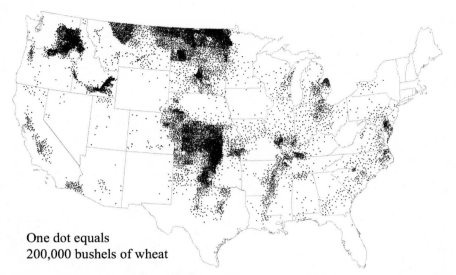

One dot equals
200,000 bushels of wheat

Figure 4.3   Wheat production, 2012. *Source*: Created by authors, data from USDA Census of Agriculture 2012.

Two massive wheat-producing districts thus emerged by the 1920s: a spring wheat region in North Dakota, which expanded westward in later years into Montana; and a winter wheat region in Kansas, Nebraska, Colorado, Oklahoma, and the Texas Panhandle (Figure 4.3). Wheat farming continued its westward movement with the frontier throughout the late nineteenth century and into the twentieth. By 1900 another wheat region had emerged south of Spokane, Washington, in the regional known as the Palouse, where rolling hills, volcanic soils, and winter-maximum precipitation proved ideal for the production of winter wheat (Meinig 1968).

These three regions have persisted for more than a century in their respective agricultural roles. Palouse winter wheat is either exported from the Pacific Coast, most often to Asian countries, or is manufactured into a variety of baking-type flours for domestic sale. The spring wheat region, which includes North Dakota and northeastern Montana, also extends well north of the border into Canada. Barley and durum wheat are both produced in the northern spring wheat region as well. In many years the barley, durum, spring wheat, and canola crops grown in Alberta, Saskatchewan, and Manitoba are larger than those produced south of the Canadian border.

Most of the rest of the wheat map shows the production of winter wheat. Outside of the major producing areas, wheat is often sown as a nurse crop for slower-growing pasture grasses, giving them time to emerge, or simply as a crop to have in the ground as a protection against soil erosion. Microbial activity in the soil is reduced when no crop is growing, so farmers sow a wheat crop in otherwise fallow ground to later plow under as "green manure." It is a desirable practice that also increases soil organic-matter content.

AMERICAN WHEAT ON THE WORLD MARKET

Wheat harvests in the United States now exceed two billion bushels in most years, but between 40% and 50% of that large harvest does not move to domestic flour mills (USDA 2011). Rather it is shipped by truck, rail, and barge to seaports where it is loaded for export around the world. Both production and exports vary considerably from year to year. This is partly due to federal price-support programs that aim to control wheat supply, and also because of fluctuations in the value of the American dollar which make U.S. products more or less attractive in price to foreign customers.

In 1915 the United States produced over one billion bushels of wheat for the first time. In the World War I years that followed, when European wheat supplies were sometimes prevented from reaching outside markets, wheat from the United States began moving abroad more often. But production and exports stagnated thereafter, through the 1920s and 1930s, as the American nation turned away from foreign alliances and commerce. World War II changed that situation abruptly and the United States soon had a new role as a wheat exporter. The growth in the American wheat crop since 1950 is tied more to the export market than it is to any growth in domestic consumption. Wheat exports increased steadily from the 1940s onward, averaging more than a billion bushels per year. Exports peaked again in the 1970s, when massive grain sales were made to the former Soviet Union.

The United States produces only about 10% of the world's wheat, but it accounts for roughly one-fourth of the world's wheat exports. Improvements in production technology and changing market forces have led to greater wheat production in Russia, Ukraine, and Kazakhstan in recent years, while U.S. production has remained constant. The former Soviet Union region is likely to export more of its wheat in the future and the United States is forecast to decline in that role (Liefert and others 2010).

Canada, Australia, France, Turkey, and Argentina also raise, store, and export large quantities of wheat. Wheat is grown in both hemispheres and on all the continents except Antarctica. The crop is harvested every six months somewhere in the world and it is produced in large surplus by more than a dozen countries. The world's food supply is not totally guaranteed by these circumstances, but the possibility of a serious food shortage is thereby reduced.

Hard Red Winter (HRW) wheat is the most widely grown variety in the United States and it is the main crop for milling into the bread flour used in

American bakeries. Only about 25% of the HRW crop is exported, mainly to Latin America and Africa (USDA 2011). In contrast, approximately half of the Durum and Hard Red Spring (HRS) wheat crops are exported. Italy, with its large pasta industry, is the largest buyer of U.S. durum wheat; Japan is the largest customer for HRS wheat. Even though North Dakota is nearly 2,000 miles from the Pacific Coast, it is the closest place producing a large quantity of bread-grade wheat that can be shipped across the Pacific. For the same reasons, Japan is often Canada's largest foreign wheat customer.

Farmers in the Palouse region of Washington, Idaho, and Oregon produce a softer white wheat that is ideal for milling into flour for the noodles consumed in Asian countries. Japan purchases 30 million bushels of American-grown soft white wheat in a typical year. The overall level of Japanese wheat purchases from the United States amounts to more than 120 million bushels per year. As a comparison, that same amount of wheat would equal the total production of North Dakota's ten largest wheat counties. Add to that the millions of bushels purchased by several dozen other countries and one quickly arrives at the sum of one billion bushels of U.S. wheat, the average annual export.

Between 10% and 30% of U.S. wheat exports are conducted through government programs providing food-aid assistance. Among the oldest of these programs is the Commodity Credit Corporation (CCC), which was created as part of President Franklin D. Roosevelt's first New Deal in the 1930s. The CCC issues commodity certificates in lieu of cash payments to program participants, meaning that farmers are paid in kind, with the crops that they offer as loan repayments. Another program, known as P.L. 480, has operated since the 1950s and allows countries to purchase American commodities with their own currency. Egypt is one of the world's largest wheat importers, consuming more than 100 million bushels of U.S. wheat in a typical year, a significant share of which is subsidized by U.S. economic assistance programs.

CONSUMER CHOICES

Flours that are whole-grain, unbleached, organic, or gluten-free add still more layers of variety to the uses made of wheat and its related grains. Educated consumers know that the flour used to make a typical loaf of white bread is bleached to make it white and is produced from only part of the kernel of wheat. The wheatberry (kernel) consists of three parts: an outer, dark-colored layer known as the bran; a starchy inner core (the endosperm); and the germ, which contains the vitamins, and is surrounded by the endosperm (California Wheat Commission 2016). In the production of white bread flour from any of the hard red wheats, the darker-colored bran and the germ are removed, the inner core is bleached, and vitamins are added to replace a portion of the nutrients that were removed. These steps are followed for the purpose of creating the white color of bread that many consumers prefer.

Hard wheats are the most desirable for manufacturing bread flour, but in the United States white-colored hard wheats were not available until a quantity of Australian white wheat was imported in the 1970s. White wheat has no major genes for bran color and its bran is also milder in flavor. In the 1990s wheat geneticists at the agricultural colleges in Kansas, North Dakota, and other wheat-producing states began developing strains of hard white wheat that could be used to produce whole-grain white bread (Paulsen 1998). The product could be made from unbleached flour that was naturally white and which had all of the nutrition of the original grain.

Some farmers planted the new hard white wheat but they encountered difficulties. Because any mixing of red and white wheat was to be strictly avoided, farmers had to have the wheat inspected and certified as 100% white. The new seed cost more than the conventional hard red wheats they had planted, the number of days to maturity of the crop was larger, and the price they received for it was not sufficient to convince large numbers of wheat farmer to switch from red to white. Supplies of white wheat were small, which forced up the price. Hard red wheat thus remained the standard when there was no incentive to switch to a new system of production.

As supplies of the new wheat slowly became available, major bread companies, including Wonder Bread and Sara Lee, began marketing a whole-grain white bread. To keep the cost down, millers used only 30% hard white and substituted hard red for the rest, thus necessitating bleaching and replacing nutrients as before. What was marketed as "whole grain" was soon revealed to be only 30% whole grain and the bakeries removed the false claims from their labels (Center for Science in the Public Interest 2008). The controversy over whole grain white bread ended without the new product gaining widespread acceptance (Lin and Vocke 2004; Vocke and others 2008).

GMO WHEAT

Still more controversial is the possibility that a genetically modified wheat will someday be grown in place of the varieties now in production. Genetically modified organisms (GMOs) are forcefully opposed by many consumer groups and environmentalists, especially in Asia and Western Europe (Berwald and others 2006). In the past, new crop varieties began with small-scale experiments at state agricultural colleges, followed by a series of crop trials, further experimentation, and then release for commercial production. Genetic engineering does not operate in that context. It is a private-sector activity, far more expensive than the older-style plant breeding, and only large biotechnology firms, such as Monsanto, Syngenta, and DuPont, can afford to do it. The companies bear all financial risks in developing a product like GMO wheat, and they move cautiously if they believe the result might not be an economic success.

GMO wheat has been in development for some time. The main purpose of the genetic modification is to make the wheat plant able to withstand applications of glyphosate herbicides (such as Roundup®, manufactured and sold by Monsanto) which kill weeds but do not destroy the crop plant. Corn and soybeans are largely GMO crops today, but wheat is not. In fact, GMO wheat is not commercially available anywhere in the world at present.

Strong, consistent opposition to GMO wheat has come from Europe and Asia. No country has been more firmly opposed to it than Japan, which abruptly canceled all of its U.S. wheat purchases for two months in 2012 after some GMO wheat was discovered growing in a farmer's field in Oregon. Other countries, including Canada, are more moderate in their stance. The potential benefits of using genetic modifications to make improvements in the wheat crop are many, they say, but Monsanto's GMO wheat might not be the most useful application of the technology (Bognar and Skogstad 2014).

The subject has remained embroiled in controversy for two decades. Monsanto was about to bring a strain of GMO wheat on the market in 1994 when a high-yielding GMO wheat was seen as a way to regain wheat acres lost to corn and soybeans. European and Asian countries reacted negatively. A feared loss of foreign markets led Canadian interests to oppose the new wheat as well. In the early 2000s the price of wheat was low and American farmers disliked the added costs of the new GMO seed.

Public statements by Monsanto in 2012 indicated that a new GMO wheat was nearly ready for introduction. The unexpected discovery of GMO wheat growing in Oregon later that year brought a torrent of public opposition once more (*Salem Statesman Journal* 2014). While the introduction of GMO wheat may be inevitable, as many say, the date when it might become available continues to get pushed forward in time.

DURUM AND BARLEY

Grain crops that are closely associated with spring and winter wheat, either historically or in terms of the uses made of them, are the grasses *Triticum durum* (durum wheat), *Hordeum vulgare* (barley), and *Oryza sativa* (rice). Durum is ground into semolina which has a high protein content and high gluten strength and is used to make pasta. North Dakota and Montana together produce nearly three-fourths of the durum grown in the United

States. Durum is grown by the same farmers who grow other types of spring wheat and its production is widespread. Fifty of North Dakota's fifty-three counties are durum producers.

It has been said that barley "has a wider ecologic niche than any other grain" (Weaver 1943). Within the United States barley is commercially produced in both Alaska and Arizona, although North Dakota, Idaho, and Montana grow about 70% of the crop. In the nineteenth century barley production moved westward much like the wheat crop did. New York was once the largest barley producer, following which Wisconsin took the lead. Barley became established on farms in Minnesota and North Dakota by 1900.

Although barley has many uses, including as a livestock feed, its primary use is for barley malt which is the principal ingredient for making beer. Malting involves steeping the barley in water and allowing the grain to sprout. It is then dried, baked, and stored before it is shipped to a brewery. Many varieties of barley are sought by beer companies looking for various colors and flavors for their brews, and they set strict specifications for malt production. Most barley farmers produce and sell their grain on contract to a maltster (malting company) which produces the malt that breweries consume (Taylor and others 2005).

Until the early 1980s the U.S. brewing industry was under the firm control of the major brand-label companies. They concentrated on brewing Pilsner or light-lager beer varieties through a process known as adjunct brewing in which non-barley starches such as rice or corn were used to supplement the malt. They purchased only a small array of barley varieties, most of which had been bred at the state agricultural experiment stations for their compatibility with adjunct brewing. But the malt varieties which the newer craft brewers desired were not those that the major-brand brewers demanded. As craft brewing began to emerge in the 1980s, those who brewed it turned to barley imports from Europe which were more suitable for making an all-malt-type product. Barley imports to the United States, which had averaged 10–15 million bushels per year, increased to 30–40 million bushels by the 1990s.

Noncraft beer production has been declining about 0.6% per year since the early 1990s but craft beer production has increased steadily to a current level of about 8% of the total U.S. beer market. Since craft beer requires between three and seven times as much malt per barrel as the noncraft types, craft beer production now consumes about 20% of U.S. malting barley production (Bond and others 2015). This has led barley breeders at the agricultural experiment stations to cooperate with the craft beer industry in creating barley varieties more suitable for craft production (O'Connell 2015). The array of barley varieties being raised for beer production has expanded as a result.

Craft beer production has a selective geography. Vermont leads all other states with 46 breweries and an annual production of 16.2 gallons per adult in the state. Colorado, Pennsylvania, Alaska, and Oregon also rank high in production. At the opposite extreme Arkansas and South Dakota produce only 0.2 gallons per adult. Nationwide there are now nearly 3,500 craft breweries, a number that includes brewpubs (bars and restaurants that use at least one-fourth of the beer they manufacture on the premises), microbreweries (producing less than 15,000 bbls/yr), and regional breweries (15,000–6,000,000 bbls/yr). Their share of the beer market continues to grow (Brewers Association 2005).

RICE

Rice is a tropical plant with origins in both Africa and Southeast Asia. The domestication of Asiatic rice (*Oryza sativa*) took place about 7,000 years ago north of the Yangtze River in China where irrigation was necessary to supply the crop's prodigious water need. It was introduced to Japan around 300 BCE. Paddy rice (wet rice), which is planted in flooded fields where water gradually soaks into the ground, is the most common rice culture in China and Japan as well as in the United States.

The Spanish introduced rice to the West Indies in the sixteenth century. The crop was being grown in the coastal American colonies by 1700 using the "skills of slaves from Madagascar and Africa" who were familiar with the crop and knew how to manage rice paddies (Sauer 1993). From there rice farming expanded west to coastal Louisiana by the early nineteenth century. Although the heritage of rice production in the American Southeast was based on African rice knowledge, the rice varieties being grown were not suited to machine processing. Japanese rice was introduced to Louisiana in 1899 and soon became the basis for the rice industry that followed (Babineaux 1967).

Farmers from the American Midwest saw the crop growing in coastal Louisiana and took it north to the alluvial Mississippi Valley where new areas of rice production were being created by 1910. Eastern Arkansas became the largest U.S. producer of rice and the farmers who grew it also organized their own marketing cooperative, Riceland Rice, which is now the world's largest marketer of milled rice.

Rice farming was established in the Central Valley of California with seed imported from Japan and China at roughly the same time. The Sacramento Valley's flat surface, flooded with irrigation water supplied by the Sacramento River, became the basis for the industry's growth. Part of the California crop is marketed through Farmers Rice Cooperative, another farmer-owned venture.

Americans do consume rice, although consumption has never been large. U.S. per capita rice consumption remained around 5 lbs/person/year until fairly recent times. The popularity of foods like sushi and other Asian special-ties, as well as growth of the Asian-heritage population of the United States, has seen an increase in per capita rice consumption to about 20 lbs. at present. But this is small compared with the 135 lbs/person/year average consumption of wheat flour.

About half of the U.S. rice crop is exported, primarily to Mexico, Central America, and Africa. Rice imports include the aromatic varieties, such as bas-mati and jasmine, which are not grown in the United States and are imported from Thailand, India, and Pakistan (USDA 2015). As is true of other grain crops, Americans produce, consume, export, and import rice for its many uses.

REFERENCES

Babineaux, L. P., Jr. 1967. *History of the Rice Industry of Southwesters Louisi-ana.* Master of Arts thesis, University of Southwestern Louisiana. [http://library.mcneese.edu/depts/archive/FTBooks/babineaux.htm].

Berwald, D., C. Carter, and G. P. Gruère. 2006. Rejecting New Technology: The Case of Genetically Modified Wheat. *American Journal of Agricultural Economics* 88(2): 432–447.

Bognar, J. and G. Skogstad. 2014. Biotechnology in North America: the United States, Canada, and Mexico. In *Handbook on Agriculture, Biotechnology, and Development,* edited by S. J. Smyth, P. W. B. Phillips, and D. Castle, 71–85. Chel-tenham, UK: Edward Elgar.

Bond, J., T. Capehart, E. Allen, and G. Kim. 2015. *Boutique Brews, Barley, and the Balance Sheet.* Feed Outlook: Special Article. FDS-15a-SA. USDA, Eco-nomic Research Service. [http://www.ers.usda.gov/publications/fds-feed-outlook/fds-15a.aspx].

Brewers Association. 2015. *Malting Barley Characteristics for Craft Brewers.* [https://www.brewersassociation.org/best-practices/malt/malting-barley-characteristics/].

California Wheat Commission. 2016. *A Kernel of Wheat.* [http://www.califor-niawheat.org/industry/diagram-of-wheat-kernel/].

Center for Science in the Public Interest. 2008. Sara Lee to Make Clear its "Made with Whole Grain White Bread" is 30 Percent Whole Grain. [http://cspinet.org/new/200807212.htm].

Kuhlman, C. B. 1929. *The Development of the Flour-milling Industry in the United States, With Special Reference to the Industry in Minneapolis.* Boston: Houghton Mifflin.

Lin, W., and G. Vocke. 2004. Hard White Wheat at a Crossroads. USDA, ERS *Elec-tronic Outlook Report.* WHS-04K-01.

Liefert, W., O. Liefert, G. Vocki, and E. Allen. 2010. Former Soviet Union Region To Play Larger Role in Meeting World Wheat Needs. *Amber Waves,* USDA Economic Research Service (online).

Malin, J. C. 1944. *Winter Wheat in the Golden Belt of Kansas*. Lawrence: University of Kansas Press.

McKelvey, B. 1949. Rochester and the Erie Canal. *Rochester History* 11(3): 1–24.

Meinig, D. W. 1968. *The Great Columbia Plain. A Historical Geography, 1805–1910*. Seattle: University of Washington Press.

O'Connell. J. 2015. Barley breeders target varieties for craft brewers. *Capital Press*. [http://www.capitalpress.com/Profit/20150312/barley-breeders-target-varieties-for-craft-brewers].

Salem Statesman Journal. November 12, 2014. "Monsanto Settles Over GMO Wheat Found in Oregon." http://www.statesmanjournal.com/story/news/2014/11/12/monsanto-settles-gmo-wheat-found-oregon/18937079/.

Paulsen, G. M. 1998. Hard White Winter Wheat for Kansas. *Keeping Up With Research*, SRL 120. Manhattan: Kansas State University Agricultural Extension Service.

Sauer, J. D. 1993. *Historical Geography of Crop Plants: A Select Roster*. Boca Raton FL: CRC Press.

Taylor, M., M. Boland, and G. Brester. 2005. *Barley Profile*. USDA, Agricultural Marketing Resource Center. [http://www.agmrc.org/commodities_products/grains_oilseeds/barley-profile/].

USDA. Economic Research Service. 2011. *Wheat Data*. [http://www.ers.usda.gov/data-products/wheat-data.aspx#25297].

USDA, Economic Research Service. 2015. *Rice Imports, Rice Exports*. [http://www.ers.usda.gov/topics/crops/rice/trade.aspx].

Vocke, G., J. C. Buzby, and H. F. Wells. 2008. Consumer Preferences Change Wheat Flour Use. *Amber Waves*. USDA Economic ResearchService (online).

Weaver, J. C. 1943. Barley in the United States: A Historical Sketch. *Geographical Review* 33(1): 56–73.

Weiss, E. and D. Zohary. 2011. The Neolithic Southwest Asian Founder Crops: Their Biology and Archaeology. *Current Anthropology* 52(54): 5237–5254.

Chapter 5

Dairy

The production of milk and the manufacture and marketing of dairy products play a central role in the food industries of the United States. National milk consumption currently averages around 200 billion pounds a year, of which fluid milk has a 30% share. Cheese, dominated by mozzarella and cheddar types, requires 11 billion pounds. Ice cream, yogurt, and butter account for much of the rest. While dairying has been an established specialty on American farms for more than 200 years, recent decades have seen major shifts in the location of dairy production and in the sizes of dairy farm operations.

THE EMERGENCE OF DAIRYING

Human use of animals for food is as old as humanity itself. Milking as a cultural practice probably began about 6,000 years ago, gradually emerging not at a single location but among separate groups of people living in various parts of the world. The nutritional benefits of milk were conferred upon those who consumed it, which may have given the milk-consuming peoples advantages that were expressed in better health and larger stature (Simoons 1974).

The milk of land mammals contains lactose (milk sugar) which is hydrolyzed by the enzyme lactase into forms readily absorbed in human digestion. Lactase activity in infants is high during the nursing period, but it declines to low levels at weaning and remains that way throughout life. Adult mammals thus have a limited ability to hydrolyze lactose, a condition known as lactose malabsorption (LM).

LM is more prevalent in some human groups than it is in others, which naturally raises the question of why this should be so (Johnson and others 1982). The most favored explanation is genetic. People who happened not to have

LM possessed a nutritional advantage that they passed along to future generations in the form of enhanced levels of health and well-being. A few thousand
years of evolution have been sufficient to create marked differences between
human groups in their ability to absorb lactose. Nearly all populations with a
low prevalence of LM today have a tradition of consuming lactose-rich dairy
products. Regions of heavy lactose absorption include northern Europe, the
Arabian Peninsula, and the northwestern portion of the Indian subcontinent.

Alternatively, most of tropical Africa and East Asia had animals suitable
for milking, but people there apparently made little use of milk for human
food. Those populations also have a higher prevalence of LM. The same is
true of Native Americans, who lacked a tradition of milking; and it is also
true of African-Americans, many of whom trace their roots to tropical and
subtropical West Africa where milking was not practiced.

While the wide dispersion of milk-consuming populations in antiquity
must have had chance beginnings, those who consumed milk would have
slowly evolved similar abilities to digest lactose-rich dairy products. Among
the groups with a low prevalence of LM are the pastoral nomads, including the Fulani of West Africa, the Hima and Tussi of East Africa, and the
Bedouin in the Arabian Peninsula (Simoons 1982).

Denmark, a country that lies in the center of a large region of low LM
in northern Europe, often leads the world in the per capita consumption of
dairy products. Other northern European populations, including the French,
Dutch, English, Germans, and Scandinavians, generally have low LM prevalence. Given the dominant place these nationalities had among immigrants
to the northern United States, it seems that the ability to digest lactose was
an incidental result of the migrations that contributed most heavily to U.S.
population growth during the nineteenth century. In some respects, the early
European-born population of the United States arrived "dairy ready."

GROWTH OF DAIRY FARMING IN THE UNITED STATES

Dairying as an agricultural specialty awaited the emergence of urban populations large enough to demand production from outside the city. By 1840 the
counties around Boston, New York, and Philadelphia had become producers
of a milk surplus that was marketed locally.

The production of fluid milk for urban markets is one stimulus to dairy
production, but there are others that are equally important. The manufacture
of butter and cheese can be carried on some distance away from population
centers. New Hampshire, Vermont, and northern New York were producing
large quantities of milk by 1840, at least part of which went into dairy products manufacturing. Distant producers in northern New England and upstate

New York could not ship milk to the city without risking spoilage, especially in the days before mechanical refrigeration. But they could convert their milk into cheese and butter, both of which store well. Dairy farming thus became a specialty both close to and far from the market, with fluid milk being supplied from nearby areas and butter and cheese from a distance, creating regions of dairy production over large areas.

By 1880 New England and New York still were the largest dairy producers, contributing more than half the total U.S. milk production. Dairy farming also moved westward, sometimes replacing wheat as a farm specialty, as the wheat frontier itself moved still farther to the west. As urban populations grew, so did the dairy industry. New York's milk production tripled from 1880 to 1925, Ohio's grew tenfold, and Wisconsin's increased more than sixty-five times in the same period. Wisconsin's dairy industries specialized in various types of cheese production, while Minnesota farmers marketed their fluid milk to local creameries which turned it into butter.

Other dairy products manufacturing specialties appeared within the areas where milk production was especially large. Evaporated milk, produced by heating milk, reducing its water content through evaporation, and then canning it, was produced in every dairy region of the United States. The introduction of mechanical refrigeration in the twentieth century led to large increases in ice-cream production. Cottage cheese, sour cream, whipping cream, and sherbets were added to the variety of products.

When the USDA issued its "Generalized Types of Farming" map in 1950, a specialization in dairy production was shown blanketing the northeastern United States, from Maine to Minnesota (USDA, 1950). Another two dozen counties in the Pacific Northwest were also designated in the dairy category. Dairying was clearly a Northern specialty, occupying the zone north of the Corn Belt and east of the Great Plains. It occupied a forested zone of abundant moisture, cold winters, and cool summers with long day-lengths (Figure 5.1).

These physical conditions were regarded as an optimal environment in which dairy cows could thrive. Leguminous hay crops, including alfalfa, clover, and lespedeza, grow quickly in the relatively short-summer but long day-length season of the Northern states. Such crops are adapted to the complex digestive system of a dairy cow which allows the animal to consume green plant material. Corn also is grown on dairy farms, but dairy farmers generally harvest the ear of corn along with the stalk and chop the mixture (known as ensilage) for storage in a silo, from which it is fed to dairy animals over the winter season (Figure 5.2).

Hay crops are essential both as bedding and for keeping dairy livestock fed, and they grow well in the northeastern United States. Two cuttings of hay, which is baled and stored where it can be kept dry, generally are possible in

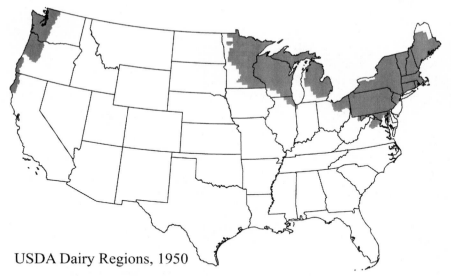

USDA Dairy Regions, 1950

Figure 5.1   Dairy region of the United States in 1950. *Source*: Created by authors, data from *Generalized Types of Farming in the United States*, 1950.

Figure 5.2 Traditional Wisconsin dairy farm with a small milking herd. *Source*: Photograph courtesy of author.

the summer season. Haying activities are aided by the short periods of dry weather that punctuate the summer season.

Added to these physical factors that favored dairying in New England, Wisconsin, and Minnesota were the cultural backgrounds of the people. The cooperative marketing of farm products began in nineteenth-century Europe and the practice quickly spread to the United States. A cooperative allows producers to achieve a price-quantity advantage in the marketplace by pooling their sales. Dairy farmers have made extensive use of cooperative marketing especially in thinly populated rural areas lacking an infrastructure for processing and marketing farm products. Some farmer-owned cooperatives have developed well-known brand labels, including Cabot cheese (owned by Vermont dairy farmers), Land o' Lakes butter (Minnesota), and Foremost dairy products (Wisconsin).

THE NEW DAIRY INDUSTRY

In 1965, when California became the most populous of the fifty states, it had already become the fourth-largest milk-producing state, behind Wisconsin, New York, and Minnesota. California's milk production was less than half of Wisconsin's at that time, when Wisconsin was unchallenged as "America's Dairyland." California's physical environment did not resemble Wisconsin's, let alone New England's. Dairying in California was split between "feedlot dairies" in the Los Angeles region, where animals were kept largely in unsheltered fashion, and still larger feedlots in the Central Valley where large herds could be kept year-round in the open air. Eventually land near Los Angeles became too expensive for use as dairy feedlots and the industry concentrated near smaller cities in the Central Valley, where it remains today. In 1993 California passed Wisconsin to become the largest milk-producing state.

The center of California's dairy industry is Tulare County, about midway between Fresno and Bakersfield in the San Joaquin Valley (Shultz 2000). Tulare County produces about $1.8 billion of milk every year, making it the top milk-producing county in the United States. The marketing and processing of milk in Tulare County follow the familiar practices instituted in New England and the Middle West more than a century ago. Dairyman's Cooperative Creamery, the largest of the farmer-owned coops in Tulare County, was purchased by the Minnesota-based Land O'Lakes cooperative in 1998. Tulare County contributes nearly one-fourth of California's total milk supply. Much of it is trucked to the Los Angeles area.

The long-held assumption that dairying was tied to the physical environment of the northeastern states, where it had traditionally been practiced, proved to be incorrect. Average July high temperatures at Wausau, Wisconsin, in the heart of the northern Wisconsin dairy region, hover around 70°, while at Hanford, in Tulare County, the July average is near 100°. Overnight low temperatures average above freezing every month of the year at Hanford, but sink below 32° five or six months of the year in Wausau. Hanford's meager 8.38″ annual precipitation—which falls almost entirely in the winter—is overwhelmed by the 32.4″ that falls in Wausau in an average year.

The lack of winter cold means that animals can be kept in open-air lots year-round in California, but a shade cover protection also is needed for the hot summer months. The Central Valley's dry climate is mitigated by the availability of irrigation water (although it, too, is restricted during severe droughts). Despite the environmental differences, farms in both Wisconsin and California feed the same crops to their dairy cows. Double-cropping of corn and wheat is practiced in California's year-long growing season. Corn, corn silage, and alfalfa are the leading crops in Tulare County's dairy industry just as they are in Wisconsin's.

The nationwide map of milk production at present reveals three distinct patterns (Figure 5.3). The traditional dairy farming region of the Northeast remains much as it has for generations, with production widely distributed

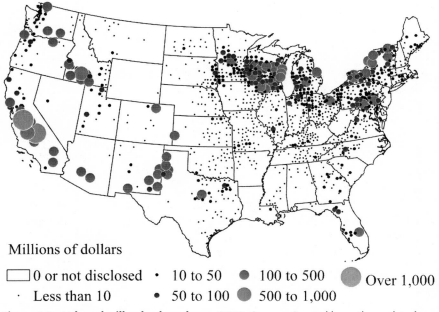

Millions of dollars

☐ 0 or not disclosed   · 10 to 50   ● 100 to 500   ⬤ Over 1,000
    · Less than 10   • 50 to 100   ⬤ 500 to 1,000

Figure 5.3   Value of milk sales from farms, 2012. *Source*: Created by authors, data from USDA Census of Agriculture 2012.

over hundreds of counties. The South is a milk-deficit region as it tradition-ally has been; Southern per capita consumption is below the national average and it relies on shipments from the Northeast and Middle West. The major shift in the national pattern is the rise of the western states' dairy industries where fewer, but much larger, operations form the pattern of production.

Along with this geographical shift has come the appearance of dairy farms much larger than were formerly thought possible. The western states' model, as first developed in California, has an average milk-cow herd size of 500–1,000. Both Idaho and Texas average between 1,000 and 2,000 milk cows per herd, which are the largest averages in the United States. All three states also have some smaller-sized dairy farms, but the small producers contribute only a tiny fraction of total production (Figure 5.4). In Wisconsin, New York, and Vermont the average herd size varies between 50 and 200 cows and the small producers dominate the industry. These older dairy states also have some of the very large dairy farms, but the large farms are a minority. The old pat-tern and the new thus coexist in the same industry, separated geographically between East and West.

The reasons for the shift to larger dairy operations involve familiar argu-ments focusing on economies of scale. Lower production costs per gallon of milk accompany larger numbers of cows per farm. Long-distance transporta-tion of bulk fluid milk is more feasible than in former times because of bet-ter milk handling and storage technologies. At the national scale, continued dispersion of the U.S. population westward and southward has demanded that products such as fresh milk be more widely available, giving rise to new patterns of transportation.

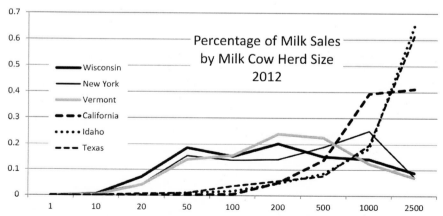

Figure 5.4   Percentage of milk sales by milk-cow herd size. *Source*: Created by authors, data from USDA Census of Agriculture 2012.

CHEESE IN THE CONTEMPORARY DAIRY INDUSTRY

Despite the appearance of dairy farms producing ten times as much milk as their smaller counterparts, both large producers and small producers continue to thrive in the same industry. Large dairies characterize the West, small- to medium-sized operations dominate in the East, but all produce a standardized product that is consumed fresh or rushed to manufacturing plants to be turned into a variety of consumer products.

The cheese industry straddles the new and old ways of doing business, especially in Wisconsin which has long ranked as the largest cheese-producing state (although California now is a close second). Modern cheese factories still receive milk directly from farms just like their predecessors did a century ago. In 2013 the state of Wisconsin had 145 operating cheese factories that were scattered over two-thirds of the state's counties (Figure 5.5). Another 163 factories cut, wrap, and shred the great wheels of cheese coming out of the cheese factories, and then label the products for name-brands such as Kraft or as house-brands for grocery-store chains. An additional 60 plants produce powdered cheese products, mostly for snack foods, and 65 more factories create processed cheese items. Adding in the 45 ice-cream factories, 12 butter creameries, 24 goat milk processors, and a list of other specialized producers brings the total to 411 dairy plants operating in Wisconsin (Wisconsin Dairy Plant Directory 2012–2013). This number includes a large Kraft Foods operation at Beaver Dam, Wisconsin, where the firm manufactures its well-known Philadelphia-brand cream cheese.

Ownership of the cheese factories is varied and far-flung. Italian-style cheese production, dominated by mozzarella cheese for pizza, is concentrated in eastern Wisconsin. Leonard Gentine and Paolo Sartori were neighbors in Plymouth, Wisconsin, in the 1940s. They founded Sargento cheese which focused on the production of Italian-type soft cheeses. Sartori later founded his own company, with the same specialization. BelGioioso, another producer of Italian cheese, is based in Green Bay. Foremost Farms, a producer of Italian and cheddar-type cheeses, was formed from two smaller Wisconsin dairy cooperatives (Golden Guernsey and Lake-to-Lake) and operates seven factories in the vicinity of Green Bay.

Swiss-type cheeses are manufactured near Monroe, in southern Wisconsin, while cheddar manufacturing is clustered in the state's more northern counties. These regional specialties are far from rigid, and many companies make more than one type, but the regional division is of long standing. Colby cheese, named for the village of Colby, Wisconsin, continues to be a specialty of that area. Blue cheese is manufactured by Lactalis, the North American division of a French company which diversified to the United States and acquired plants in both Wisconsin and Idaho to produce French-type cheeses.

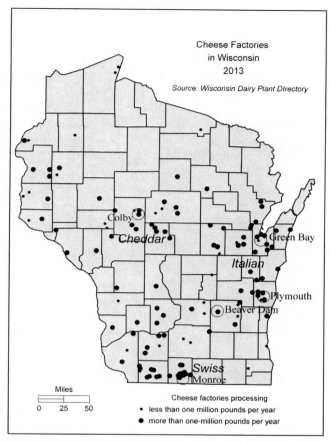

Figure 5.5   Cheese factories in Wisconsin, 2013. *Source*: Created by authors, data from Wisconsin Dairy Plant Directory.

Added to this diversity is Quebec-based Saputo, Inc., an international producer of milk, cheese, and related products. The Saputo family emigrated from Sicily to Quebec in the 1950s and now owns dozens of plants in the United States, Canada, Australia, and Argentina. Saputo plays a major role in the dairy industry of Tulare County, California, and is a leader in the production of extended-shelf-life dairy products.

MILK PRODUCTION AND CONSUMPTION

Per capita fluid milk consumption in the United States peaked at 42.3 gallons per person in 1945 and has declined steadily to around 19.6 gallons at present

(Bentley 2014). While Americans continue to consume about 8 ounces when they do drink milk, the occasions on which they drink it have become fewer in number. The only consumption trend offsetting the decline in milk drinking is the increase in cheese consumption, which has tripled on a per capita basis since 1970 largely due to the popularity of pizza. Even ice-cream production has declined about 25% since 1970 (Stewart and others 2013).

Although milk has declined in popularity as an American beverage, consumers are no less interested today than they were in the past in the various forms in which milk is made available to them. Door-to-door milk delivery was common in the United States up through the 1940s. Years before "recycling" became a household word, commercial dairies delivered milk in glass bottles which the consumer rinsed when empty and placed for pickup the next day so they could be returned to the bottling plant for sterilizing and reuse. Cream rose to the top of the whole milk and was carefully skimmed off by the customer before any milk was poured from the bottle. Cream has butterfat content between 18% and 20% by weight, whereas whole milk typically ranges between 3.25% and 5% butterfat.

After the 1940s most milk was homogenized, a physical process that reduces the butterfat globules to the size that they can be uniformly distributed through the milk. Reusable glass bottles soon were replaced by wax-coated paper cartons. Consumer concerns about butterfat intake increased the popularity of low-fat milk, which typically contains between 1.5% and 1.8% butterfat. Homogenized milk, whether whole, low-fat, or skim (less than 1% butterfat), has dominated the market for several decades. Lower-fat milk generally has vitamins A and D added to replace nutrients lost when the butterfat content is lowered. These many options and varieties are supplemented by cow's milk offered in lactose-free form, or even in entirely non-dairy products such as almond milk and soy milk.

Homogenization is relatively free of controversy among consumers, but pasteurization (a process with which it is sometimes confused) remains an issue of contention. Louis Pasteur's discoveries of the 1860s recognized that microbial growth could be inhibited in beer and wine by heating the liquid to 145°F for thirty minutes and then cooling it rapidly. Later, the process was applied to milk and various temperature levels and heating times were specified. Conventional pasteurization is a means of destroying about 90% of possible bacterial growth without resorting to the still higher temperatures needed for sterilization which change the flavor of the product.

At the opposite extreme is raw milk, which is neither pasteurized nor homogenized. Raw milk advocates point to its health benefits (a higher vitamin C content in some cases), while its detractors note that drinking unpasteurized milk risks contracting a variety of infections and diseases, including tuberculosis. Raw milk is widely sold in Europe, but in the United States retail sale is legal only in ten states. Raw milk is totally illegal in eleven states and its sale is variously restricted in all of the rest (National Association of State Legislatures 2015). The two largest U.S. organic milk producers, Horizon and Organic Valley, do not sell raw milk and both use what has been termed "ultrapasteurization"—heating the milk very briefly to 280°F, then cooling it rapidly. This Ultra-High Temperature (UHT) pasteurization kills all bacteria in the milk, permitting it to be stored without refrigeration as long as it remains unopened. Critics of UHT pasteurization note that the higher temperatures adversely affect the milk's flavor.

While the raw milk issue continues to divide organic food advocates, most groups are firm in their opposition to the so-called Bovine Growth Hormone (BGH). Bovine somatotropin (BST) is a natural enzyme produced in cows. The genetic engineering firm Genentech developed a gene for BST in the 1970s for the purpose of increasing milk production in dairy cows. The product was licensed for commercial production by Monsanto. The U. S. Food and Drug Administration approved BGH for use in 1993 and reconfirmed their findings in 1999 (U.S. Food and Drug Administration 2015).

Canada, Australia, Japan, and the EU banned BGH and consumer groups in the United States continued their opposition as well. Although BGH was widely adopted at first, it gained in unpopularity and drew negative consumer reactions. Monsanto sold its BGH product line to the Eli Lilly pharmaceutical firm in 2008. Several U.S. grocery-store chains, including Safeway and Kroger, have pledged not to sell milk from cows injected with BGH.

MILK GRADES

For many years the farms that produced milk for sale were grouped into two classes based on the quality standards their product could meet. Fluid milk—the kind that people drink—had to meet the Grade A standard, while milk sent for manufacture into butter and cheese had a lower standard, known as Grade B. Scattered dairy farmers on the fringes of the major dairy regions were Grade B producers. They were far from the market and sold their milk to local creameries or cheese factories.

Over time the standards for Grade A and Grade B production have been brought closer together, and the proportion of Grade B producers has dropped (Chite 1991). By 1991, 92% of U.S. milk came from Grade A producers, although Wisconsin and Minnesota still had more than 16,000 Grade B producers who mostly supplied the butter and cheese factories of those two states. Given the disappearance of Grade B production in recent years, however, most butter and cheese products are now obviously made from the same Grade A milk that consumers drink.

The prices paid for Grade A and Grade B milk are set by administrative rule but are not greatly different. Besides a lack of price incentive, other reasons that farmers might not convert to the Grade A standard include the requirement for a water supply that meets drinking-water standards and the purchase of expensive refrigeration equipment, both mandated for Grade A production. In Wisconsin Amish dairy farmers, many of whom do not use electricity and therefore cannot use mechanical refrigeration devices, have continued as Grade B producers (Cross 2014). But Grade A farms now account for 96% of milk production in Wisconsin.

In California, 98.4% of producers were Grade A in 2013, although the state still produced several hundred million pounds of Grade B milk a year, mostly in Merced County, 100 miles north of Tulare. Merced is also second to Tulare County in overall milk production. Like Wisconsin, California uses Grade B milk for products like mozzarella cheese, which California leads the nation in making (California Dairy Statistics Annual 2013). Grade B milk thus lives on, although its share of total production continues to shrink.

REFERENCES

Bentley, J. 2014. Trends in U.S. per Capita Consumption of Dairy Products, 1970-2013. USDA, Economic Research Service [http://www.ers.usda.gov/amber-waves/2014-june/trends-in-us-per-capita-consumption-of-dairy-products,-1970-2012.aspx#. VlTMAnarSUk].

California Dairy Statistics Annual. 2013. Sacramento: California Department of Food and Agriculture.

Chite, R. M. 1991. *Milk Standards: Grade A vs. Grade B.* Congressional Research Service, CRS Report for Congress. Washington, DC: Library of Congress.

Cross, J. A. 2014. Continuity and Change: Amish Dairy Farming in Wisconsin over the Past Decade. *Geographical Review* 104(1): 52–70.

Generalized Types of Farming in the United States. 1950. Agricultural Information Bulletin No. 3. Washington: U. S. Department of Agriculture.

Johnson, J. D., N. Kretchmer, and F. J. Simoons. 1982. Lactose Malabsorption: Its Biology and History. *Advances in Pediatrics* 21(1): 197–237.

National Association of State Legislatures. 2015. *State Milk Laws.* [http://www.ncsl. org/research/agriculture-and-rural-development/raw-milk-2012.aspx].

Shultz, T. 2000. *The Dairy Industry in Tulare County.* University of California Cooperative Extension.

Simoons, F. J. 1974. The Determinants of Dairying and Milk Use in the Old World: Ecological, Physiological, and Cultural. In *Food, Ecology, and Culture: Readings in the Anthropology of Dietary Practices,* edited by J. R. K. Robson. New York: Gordon and Breach.

_____. 1982. Problems in the Use of Animal Products as Human Food: Some Ethnographical and Historical Problems. In *Animal Products in Human Nutrition*, edited by D. C. Beitz and R. G. Hansen, 19–34. New York: Academic Press.

Stewart, H., D. Dong, and A. Carlson. 2013. *Why Are Americans Consuming Less Fluid Milk?* USDA, Economic Research Service. Economic Research Report 149.

U.S. Food and Drug Administration. 2015. *Report on the Food and Drug Administration's Review of Recombinant Bovine Somatotropin.* [http://www.fda.gov/Animal-Veterinary/SafetyHealth/ProductSafetyInformation/ucm130321.html].

Wisconsin Dairy Plant Directory, 2012-2013. Madison: Wisconsin Department of Agriculture, Trade & Consumer Protection.

Chapter 6

Pork and Beef

Despite their scattered origins over various parts of Eurasia, as well as differences in their sizes, diets, and habits, hogs and beef cattle were raised together, often in adjoining lots, and fulfilled complementary roles on farms of the American Corn Belt for nearly 200 years. It was not until factory-type methods began to appear in the 1970s that pork and beef production began to separate functionally as well as geographically. Before that time, where there were cattle there were also hogs, and both were fed the corn that was harvested on the farms where the animals were raised. Those same farms also kept horses, had flocks of chickens, and sometimes raised sheep and lambs as well, but such animals were subject to separate treatment because of their different needs.

THE HOG

No large animal has provided more for human sustenance yet has demanded less than the hog. Like cattle, the arrival of hogs in the Western Hemisphere involved a mixture of care and neglect on the part of the European colonizers who brought them. The first eight hogs known to have set foot on the Caribbean islands came with Columbus in 1493, but the date has little significance because nearly every ship that followed brought more. If hogs had been in need of human care, they likely would not have survived at all because little attention was paid to them except when it was time to butcher.

Hogs, like cattle, are quite capable of living on their own. Cattle, hogs, and horses that were survivors of shipwrecks took up residence on coastal littorals and islands of North America from the sixteenth century onward. The hogs became the razorbacks of the southeastern woodlands of the United States.

Their ability to go from domesticated to feral condition, survive for generations, and then come back to a domesticated existence partly accounts for
their great value as a frontier animal. Hogs were not only a source of meat;
they also provided hides for leather goods, lard oil for heating, and grease for
making soap.

Hogs are members of the family *Suidae*, which means they are four-legged
ungulates, have four toes on each foot, and possess a simple stomach (rather
than the more elaborate digestive system of cows). Dozens of extinct genera
of Suids are known from fossil records extending from Asia Minor to the
islands of Indonesia and the Philippines. Wart hogs and bearded hogs, ranging from Southeast Asia to northern Africa, are close relatives.

The hogs raised on American farms are descended from the European wild
boar, *Sus scrofa*, which shows signs of having separated from the ancestors of
other *Sus* species about 500,000 years ago (Giuffra and others 2000). Domestication of hogs did not take place until about 9,000 years ago in Europe and
perhaps 1,000 years after that in China, which means there was a long period
of evolution among animals in widely separated areas. The swine that Columbus, DeSoto, and the other Spanish explorers brought with them were semi-
domesticated animals that had interbred freely with wild boars in areas such
as Extremadura on the Spanish-Portuguese borderlands. The animals were
used to being driven by swineherds and could be penned or allowed to roam.

Small numbers were taken aboard ships for purposes of food supply on the
long sea voyage to the New World. Those that survived sometimes escaped
and ran into the forest as soon as the ship had reached land. Repetition of
this sequence for a century or more created a supply of feral hogs around the
southeastern coast of the United States that was free for the taking. DeSoto's
expedition of 1539 observed that native people in Florida were raising a
small, barkless variety of dog as table food. Once hogs were introduced, the
natives appear to have adopted pork eating immediately and the dog was no
longer raised (Sauer 1971).

The razorback hog that was present in North America from the first European arrivals until the late eighteenth century was derived entirely from the
European wild boar. The animal had the long legs, long snout, sharp back,
and roaming instinct of its European ancestors (Dawson 1913). Improvement of some of those traits awaited the importation of domestic stock from
Europe. Asian pigs, which had descended from the same wild species also
found in Europe, had become somewhat smaller than the razorback, had
shorter legs, and provided more meat that was also of better quality. Pork
was commonly available in China just as it was in Europe, and Asian pigs
were sent from China to northwestern Europe for breeding purposes in the
eighteenth century. When importations of European hogs were made to the
United States around 1800, the Asian traits had already introgressed into

those of the European hogs. From that time onward, the domestic hog was an admixture of European and Asian traits.

American hog breeders soon began experimenting to produce a better animal. One result was the Chester White, which originated in Chester County, Pennsylvania, where it had been bred from stock imported from Bedfordshire, England, and which incorporated many of the desirable traits of the Asian animal. Massachusetts, New Jersey, and Maryland farmers experimented in similar fashion and created breeds such as the Berkshire, Duroc, and Jersey. By 1800, when the movement of settlers into the Ohio Valley was well under way, improved hog breeds were widespread in their influence (Hudson 1994).

In 1816 the Union Shaker community of Warren County, in the Miami Valley of Ohio, brought a boar and three sows of another English breed, the Big China, from Pennsylvania. When crossed with the Berkshire and another breed, the Irish Grazier, the result was an animal that walked to market without difficulty and produced rapid weight gain on limited feed and care (Davis and Duncan 1921). It became known as the Poland-China breed of swine and has persisted to this day as one of the most favored hog breeds of the Corn Belt.

The pivotal role played by swine in the development and growth of the Corn Belt relates as much to the corn crop as it does to the animal. Hogs can eat and digest nearly anything they find, which is one key to their survival in a feral state. Corn was not part of the hog's diet until the animal arrived in the American colonies where corn was already being raised. Early American hogs fed on nuts, acorns, and roots—collectively known as the woodland mast—which was their typical diet until well into the eighteenth century, just as it had been in Europe. But corn fattening produced heavier animals more quickly and it resulted in a superior quality of meat and lard. The availability of a corn crop made pork production a more lucrative business just as the steady demands of the hungry animals stimulated increased corn production. Hog production and corn production thus grew to become practically coextensive, and the result, of course, was the Corn Belt, the massive agricultural region that stretches from Ohio to Nebraska.

BEEF CATTLE

The ancestry of the cattle breeds now present in North and South America was influenced by the same processes of European overseas exploration and migration that created the early establishment of hogs. All domestic cattle are members of the genus *Bos*, which is one of about a dozen genera of the bovid family (subfamily *Bovinae*). Cattle are natural grass-eating animals. They have long tongues for twisting grassy forage, large teeth for chewing it, and a four-chambered stomach for digesting it. Their common ancestor was the aurochs (*Bos primigenius*), which ranged widely over Europe and Asia Minor until the last recorded individual died in Poland in 1627. Just as hogs were primarily associated with woodland environments, cattle were at home in open, grassy landscapes (Trow-Smith 1957).

The wild aurochsen spread across the full extent of Europe, Asia, and North Africa, developing genetic variations over time. At least two independent domestications took place, one in Anatolia and the other in the Indian subcontinent, about 10,500 years ago. From that time onward two domesticated species are recognized. *Bos taurus* (or taurine cattle), concentrated in Asia Minor and Europe; and *Bos indicus* (or indicine cattle), concentrated in India and Pakistan (Felius 1985). Modern genetic studies suggest that both types influenced the cattle of North Africa (McTavish and others 2013).

Taurine cattle were typically horned, straight-eared, and large bodied, while indicine cattle were smaller in size, usually horned, lop-eared, and were most easily distinguished because of the pronounced hump on their backs. Among the breeds of indicine cattle the Indian zebu is well known for its ability to endure hot climates and resist various tropical diseases. Taurine cattle, which are most closely associated with northern Europe, lack those traits. Because cattle with indicine traits were present in North Africa, it is possible that the Moors brought this influence with them when they arrived in Spain and Sicily in the eighth century.

The first cattle that came to the Western Hemisphere were brought by Columbus in 1493. They were taken aboard in the Canary Islands and released on the island of Hispaniola. The cattle brought by the Spanish were primarily of the taurine type, although they had a small amount of indicine ancestry. On Hispaniola and other Caribbean islands and in Mexico the cattle fended for themselves under survival-of-the-fittest circumstances for nearly 400 years. On their own they essentially created their own breed, which became known in the United States as the Texas Longhorn. Resistance to tropical diseases and the ability to withstand both heat and drought set the Longhorn distinctly apart from the cattle brought to the eastern seaboard by the English colonists (Rouse 1977).

Horned cattle obviously possess a defensive advantage in situations where animals must fend for themselves and where there is plenty of space in which to roam. Cattle without horns are termed "polled," but they do not represent a separate species. Whether an individual animal has horns or not is up to simple genetic inheritance. Although an animal's horns can be physically removed, the procedure does no good for the animal. Deliberate selection probably favored polled cattle in the colder regions of Europe where animals had to be sheltered in a small space in winter. Polled, rather than horned, cattle became the preference in northern Europe by the 1700s.

Efforts at cattle breeding in both Great Britain and the Netherlands produced animals with distinct advantages in either beef or dairy production but rarely in both. The two breeds that achieved the greatest recognition as beef cattle were the Angus, a purely black species bred in Aberdeenshire and Angus in northeast Scotland; and the Hereford, a red-and-white breed that originated in Herefordshire, England, on the border with Wales (Sanders 1928; Heath-Agnew 1983). Aberdeen Angus and Hereford (Whiteface) cattle imports were made to the United States as soon as the Revolutionary War had ended and trade was resumed. They became the principal breeds for upgrading the native stock kept on American farms (Briggs and Briggs 1980).

CORN FEEDING

Corn (maize) is a New World crop just as certainly as cattle originated in the Old World. The revolution that took place in the Appalachians and the Ohio Valley at the end of the eighteenth century was to combine these traditions by switching the grazing animals' diet from grass to corn, and to feed intensively on corn for three to four months before driving the stock to market. The cattle gained extra weight, the meat they produced was of a preferred quality, and the animal brought more in return at sale than had been the case before. A considerable amount of corn passes whole through a corn-fed steer but hogs clean up this waste, literally "behind" cattle in the feedlots.

Cattle feeding was invented by a small group of farmers who kept stock in the western valleys of Virginia during the late eighteenth century. The animals were penned in open lots outdoors rather than in barns. Each lot typically contained 100 head of cattle in the fattening season and had an assortment of hogs as well. In late winter each lot of cattle and hogs was driven to market together, to Baltimore, Philadelphia, and other eastern markets. By 1800 the pioneer Virginia cattlemen had found a larger supply of better land in the Scioto Valley of Ohio. They moved there and established a cattle feeding industry which was also based on livestock droving to eastern markets as it had been in the past (Henlein 1959). By 1880 this style of cattle raising had spread from the Scioto Valley west across Indiana, Illinois, and Iowa, and into Nebraska and Kansas (Figure 6.1).

The horned cattle that the Spanish bought to the Caribbean islands and Mexico had remained largely untended and were left to multiply in number. They were overrunning the rangelands of South Texas by the middle of the nineteenth century. The legendary cattle drives characterizing this period in the history of the Great Plains originated with efforts to direct Longhorns northward to railway lines where they could be carried east to market. Unlike the European-derived stock then being raised in the Corn Belt, Longhorns

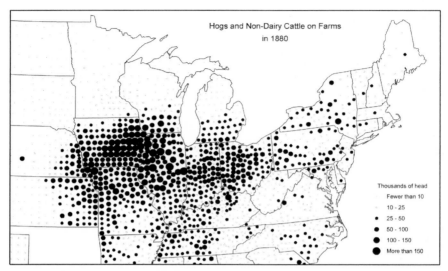

Figure 6.1   Hogs and non-dairy cattle on farms, 1880. *Source*: Created by authors, data from USDA Census of Agriculture.

could not gain weight rapidly. They were not suited to grain fattening, but when crossed with the Hereford blood line Longhorn cows produced off-spring that were good feeders. By the late 1880s Longhorns had nearly been bred out of existence.

PORK AND BEEF AS THE CORN BELT STANDARD

The concentration of hogs and non-dairy cattle that blanketed the Corn Belt by 1880 has remained the heart of the region ever since that time. Lands south of the Corn Belt supported livestock, but corn production was limited by poorer soils and by hotter weather and shorter day-lengths in the summer growing season. Expansion to the north would await the development of shorter season hybrid varieties of corn which eventually attracted livestock production northward as well. In 1880 the Plains states had just started receiving agricultural settlers from the Middle West, but these states, too, would fill with corn and livestock producers in a few decades.

The meat packing industry grew in the Middle West as livestock production advanced to the west. Gustavus F. Swift and Philip D. Armour were the leaders in building Chicago's packing industries and they helped develop similar interests in other Midwestern cities as well. The Chicago packers organized their businesses around the receipt of live animals at the city's

Union Stockyards. Other packers in or near Chicago, including Thomas Wilson, Patrick Cudahy, and Nelson Morris, also built massive meat-processing facilities around the yards. Statistics in the Census of 1890 show that Chicago accounted for nearly half of the urban wholesale meat business of the United States.

An industry that large was bound to grow and expand, especially to the West where an increasing share of the animals originated. By 1910 Illinois packers produced only about one-fourth of the beef and pork supply, while nearly a third came from packing plants in Kansas, Missouri, and Nebraska. Those more western packing plants were branch operations of the Chicago packers in most instances, however, and Chicago thus remained at the top of the industry even as East St. Louis, Omaha, and Kansas City grew. Still more branch operations by the Chicago packers soon appeared in St. Paul, St. Joseph, Indianapolis, and Sioux City.

The cattle business was organized like a large funnel that opened to the West. Cattle were born and lived their first year or two on the grassy ranges of the Dakotas, Nebraska, Wyoming, and Montana. Large ranches still characterize that region today, as they do the drier interior portions of Oregon, California, and Nevada. Cattle auctions in the western Great Plains and Intermountain West region attract ranchers with stock to sell as well as buyers from the corn-raising Middle West who purchase young stock for fattening on corn. Large herds of beef cattle grazing on large farms and ranches characterize much of the western Great Plains.

CHANGES IN THE MEATPACKING INDUSTRY

Up through the 1950s, animals were born on the western range, fed to market weight on Corn Belt farms, and then shipped again to one of the Union Stock Yards in the likes of Chicago, Omaha, Kansas City, Sioux City, or St. Paul. Cattle and hogs were purchased there by agents of the meat packers and the animals then walked their last steps to the shambles. In the 1960s this long-standing stockyards-packing plant arrangement began to fall victim to various changes in the meat industry. Competing for the supply of high-quality animals, meat packers began sending their cattle buyers directly to the feedlots, bypassing the big-city stockyards. As the volume of cattle handled by the stockyards declined, their once-important role ceased and by the late 1990s big city stockyards were only a memory.

The concentration of the meat industry into scarcely more than a dozen cities of the Middle West required an extensive network of cold-storage warehouses. Owned by the major packers, they were scattered over the nation's cities as wholesale distribution centers. Sides of dressed beef and whole hogs were loaded into refrigerated railroad cars at the packing plants and sent to the many warehouses. The railroad cars were unloaded by hand at the warehouses and the meat was prepared for sale using truck delivery to local grocery stores.

In the 1960s two small packing firms at Dennison and Fort Dodge, Iowa, consolidated under the name Iowa Beef Packers and began marketing their product in the manner that became known as boxed beef. In the old system about one-sixth of the carcass, which was inedible fat and bone, was removed and discarded at the wholesale warehouses where meat was prepared for sale. By removing this unsalable portion and wrapping the product for supermarket delivery before it left the packing plant it was possible for refrigerated trucks to accomplish all of the transportation, delivering the product straight from the packing plant to the supermarket.

As boxed beef gained acceptance after 1970, new packing plants were built as close to the supply of fed cattle as possible. This meant the closure of long-standing operations in the likes of Chicago, Omaha, and Kansas City and the creation of new packing facilities in a scatter of cities and towns, including Grand Island, Nebraska; Garden City, Kansas; Greeley, Colorado; and Amarillo, Texas. The new meatpacking centers lacked a labor force large enough to support a packing plant which led to the controversial practice of importing laborers from countries such as Mexico and Vietnam. Bitter disputes over labor exploitation, mistreatment of ethnic minorities, and challenges to community traditions followed these changes and have not diminished in some meatpacking centers down to present times (Broadway 2007).

A third major shift affecting the beef industry was the necessity to supply feed grains such as a corn and sorghum to the new feedlots in the dry Great Plains where such crops have always been marginal without irrigation. The High Plains aquifer was tapped for deep-well pump irrigation, but the water supply has never been sufficient to raise all the grain consumed by the millions of cattle fed in the western Plains feedlots. Corn is shipped in by rail from Illinois and Iowa farms to augment what can be grown with the aid of irrigation in the High Plains.

The many changes that took place in the U.S. beef industry after the 1960s had other impacts as well. One outcome was the removal of the long-standing connection between beef cattle and hogs. No hogs "follow" cattle in the feedlots as they once did. Today's Corn Belt farm family is more likely to sell their crops for cash than to feed them to animals on their own farm because their farm is a cash grain operation that has no livestock. On the business end, corporate giants of meatpacking, such as Armour, Swift, and Cudahy, did not survive the tumultuous changes that took place. Nearly all of them failed financially, were reorganized as units of larger corporations, and essentially disappeared as businesses. But their brand labels have retained consumer respect, surviving the companies that created them. John Morell, Armour-Eckrich, and Swift Premium products, to name a few, can be found in the grocery store's meat section today, even though those companies no longer exist.

GRASSFED VERSUS GRAIN-FED CATTLE

The episodes of corporate failure and financial mismanagement that took place after 1960 may have contributed to a growing mistrust of other meat industry practices such as the grain feeding of cattle. Although fattening cattle on corn in open-air lots was a practice that originated in late eighteenth-century Virginia, and was almost universally practiced throughout the Corn Belt thereafter, grain fattening began drawing negative commentary from various sources about the time boxed beef was introduced.

The fact that all cattle are natural grass-eating animals can be used to argue that they should eat nothing else, but the terms "grass" and "grain" do not denote different types of plants. Corn is a grass, a member of the *Poaceae* family, yet it is the principal feed grain used to fatten cattle. Alfalfa, which is one of the most prized forage crops for cattle, is a member of the *Fabaceae* (*Leguminosae*), a group of flowering plants that does not include the grasses. As interest in grassfed beef grew, a variety of parties organized the American Grassfed Association (AGA) at Denver, Colorado, in 2003 (American Grassfed Association 2015). One of their first tasks was to provide an unambiguous definition of what it means for cattle to be "grassfed."

The AGA and other groups persuaded the USDA to formulate a grassfed standard to be used in advertising and marketing beef. The first standard proposed by the USDA was to certify animals as grassfed if their diet after weaning was 80% grass (USDA 2002). Pressure from interested parties led to increases in the standard until it reached 100% in 2008 (USDA 2009). The USDA rule reads, in part: "the diet shall be derived solely from forage consisting of grass (annual and perennial), forbs (e.g. legumes, *Brassica*), browse, or cereal grain crops in the vegetative (pre-grain) state." For its own purposes, the AGA added three stipulations to the USDA standard: animals must be raised on pasture without confinement to feedlots, they are never to be treated with antibiotics or growth hormones, and "all animals are born and raised on American family farms." The USDA grass fed standard was in force from 2006 to January 2016, when the Agricultural Marketing Service rescinded it because of conflicts with another USDA agency, the Food Safety Inspection Service, over how to administer it (USDA 2016).

The roughly 225 members of the AGA are scattered widely in the United States, with several obvious clusters (Figure 6.2). Central Texas, where Longhorn cattle once roamed by the hundreds of thousands, has the greatest

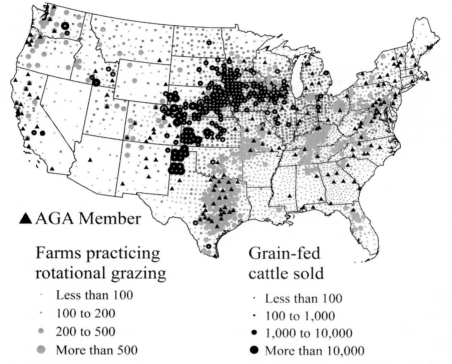

▲ AGA Member

Farms practicing   Grain-fed
rotational grazing   cattle sold

  ·   Less than 100   ·   Less than 100
  ·   100 to 200   ·   100 to 1,000
  ◦   200 to 500   ●   1,000 to 10,000
  ◉   More than 500   ⬤   More than 10,000

Figure 6.2   Grassfed cattle, grain-fed cattle, and members of the American Grassfed Association, 2012. *Source:* Created by authors, data from USDA Census of Agriculture and American Grassfed Association.

concentration of AGA members. Colorado, California, and the northeastern states are home to many members as well. While not all AGA members produce beef for sale (sheep and lambs are also included), that is the main business with which the association is concerned.

The U.S. Census of Agriculture has long published statistics on grain-fed cattle numbers and sales. Animals that are only grass fed are not the subject of a similar set of published statistics, but the practice of grass feeding can be roughly approximated with a proxy. Rotational grazing involves the frequent rotation of small groups of beef cattle from one portion of a larger pasture to another which allows time for forage to recover and lets the animals remain in fresh grass at all times. It is regarded as an improvement over "straight" grazing, where cattle graze the same ground for long periods of time.

Rotational grazing is seen as a valuable assist in producing grassfed beef. It is practiced on many farms that flank the lower slopes of the Appalachian Mountains in the east, where farms are small and animals are few

(Hart 2007). Rotational grazing also dominates in a swath of counties from central Missouri, through the Ozarks, across Oklahoma, and down the length of the state of Texas. Grain fattening, which has a long-established region of dominance from northwest Iowa south through the irrigated High Plains to the Amarillo, Texas, vicinity is equally obvious on the same map. The grass-fed and grain-fed belts are separated by the Great Plains winter wheat belt, stretching from central Kansas to the Texas Panhandle.

What accounts for the clearly separate regions of grainfed versus grassfed beef production? Comparison with a map of corn production probably gives the best clue. Corn is not a productive crop in the counties where rotational grazing is practiced most intensively. Poorer soils, hilly land, and dry conditions limit corn production but none of those limitations seriously challenge the growth of pasture grasses. Grass feeding can thus be viewed as an expedient in those areas where corn crops are too poor to support a cattle industry based on grain feeding.

CAFOs AND CONTRACTORS

Little has remained constant in the U.S. pork industry over the past fifty years, although the largest share of the hog crop is still born, fed, and slaughtered in the Middle West. Per capita consumption of pork stands at around 50 pounds per person today, down from a long-term average of 60 pounds. Apart from those indicators of relative stability, the pork business has undergone massive changes and it has been the focus of intense criticism from many sources.

Negative reaction to the operation of CAFOs (confined animal feeding operations) probably has received the greatest media coverage. Consumer groups, animal rights activists, and environmentalists have opposed CAFOs on several grounds:

Pollution: CAFOs produce solid waste in large, concentrated amounts. The waste cannot be disposed of easily and it gives off noxious fumes that foul the air. Ponds for holding the waste sometimes leak which creates spills into rivers and causes contamination of groundwater.

Treatment of animals: CAFOs treat animals in an inhumane fashion by confining them to a small space and by controlling their movements. Animals may be injected with antibiotics that are harmful to humans when the product is consumed. The massing of large numbers of animals in confined spaces is itself a dangerous practice for health reasons.

Fairness: CAFOs are large farms and they set unfair competitive conditions for small farmers. They are sometimes owned by corporations as part of a strategy of vertical integration which is also unfair to family farmers. State laws have encouraged vertical integration ("factory farming") with so-called "right to farm" laws in some states while other states have passed laws severely restricting the practice.

Balancing some of the negative arguments are issues such as improving animal health and increasing meat quality that led to the creation of CAFOs in the first place. "Old- fashioned" hog raising was based on the proverbial pig pen, where muddy hogs lived in sloppy conditions and were fed a diet of ear corn and whatever else may have been at hand. Any attempts to improve the animals' health, curtail the spread of disease, or improve the overall quality of the product were doomed by those conditions. Concrete floors were introduced in the 1960s, but they were harmful to the feet and legs of any animal that had to stand on them. Then came floors with a rubberized coating, partially enclosed feeding areas, and the imposition of strict rules limiting access by outsiders. This trend almost inevitably led to moving the entire operation indoors (Figure 6.3). By the 1990s hogs were disappearing into long sheds resembling poultry houses, where they were further isolated from disease vectors and kept at a controlled temperature. The sight of hogs wallowing in the mud became yet another aspect of farm life relegated to the story books.

Figure 6.3   Hogs in a present-day confinement barn. *Source*: Photograph courtesy of author.

As the production of meat came under the increasing influence of grocery-store chains packers began to focus on several principles that retailers demanded:

- the product must be uniform in quality, color, flavor, and texture.
- it must pass the highest possible standards of healthfulness.
- it must be available over the entire year.
- its presence on a given store's shelves on a given day has to be assured.

Imposition of these standards by one meat producer obviously led to the adoption of similar standards by all others as well.

Once the drive toward product standardization and safety was under way the inherent economies of scale in hog raising became evident. The way was clear to create hog feeding operations of colossal size, surpassing anything that had been known before. In the 1960s, Wendell Murphy, a hog farmer in eastern North Carolina, began to organize his hog production along these lines (Hart 2003). Murphy concentrated more than twenty times as many animals in a single operation than had been the practice in times past. He constructed a grain elevator to store feed grain imported by rail from the Midwest, and built what became the model of the CAFO hog farm. Murphy was also a North Carolina state legislator and he helped influence changes in state laws that would shelter CAFOs from the force of anti-pollution rules that were increasingly being enforced by the U.S. Environmental Protection Agency. Murphy became a lightning rod attracting anti-CAFO criticism.

A flurry of other changes accompanied the rise of the new style of feeding operations. Since the early days of hog farming, animals had been owned by the farmer who fed them until they were sold to a stock buyer or on the open market. Major meat packers wished to control their sources of animals more completely than that, however, and they began using a variation of the production contract arrangement that had originally been applied in the poultry industry.

Today nearly all hogs processed by the major packers are owned by the company from embryo to birth to slaughter (Harper 2009). The hogs are the product of controlled genetics, they are born in specialized swine breeding facilities, and they are delivered to individual farmer-feeders who will raise them. The intermediary role is performed by a contractor or integrator who delivers the young pigs, provides their feed, and extends various production services to the farmers. The animals are sold at a predetermined weight that the meat processor desires.

Hogs still are raised on family-owned and -operated farms, but the farmer who feeds the animals to market weight typically no longer owns them. Vertical integration, which had long been regarded as impractical in American agriculture, took over the production of hogs in the 1990s. Product standardization can again be identified as the driving force behind the change. Producing a consistently high-quality product demands a high degree of conformity when many hands are doing the work. Accountability is necessary in such a system as well, such as when a health issue is raised and a product recall is instituted. The tightly integrated, birth-to-slaughter production of meat animals is under the control of a few very large companies that are now, increasingly, under foreign ownership. Those critical of this system often cite the uniformity and lack of variety in the products that has resulted, although that, of course, was a central objective of those who established the system in the first place.

Wendell Murphy's organization of Murphy Farms soon led him into the meatpacking business. Murphy's area in eastern North Carolina had become the second-largest concentration of hog raising in the United States (after northern Iowa/southern Minnesota) by the 1990s. Other entrepreneurs, seeing Murphy's success, launched similar ventures. Seaboard Corporation, an international food, transportation, and energy company, constructed a large meatpacking plant at Guymon in the western Panhandle of Oklahoma (Hart and Mayda 1997). The company contracted hogs with local producers and

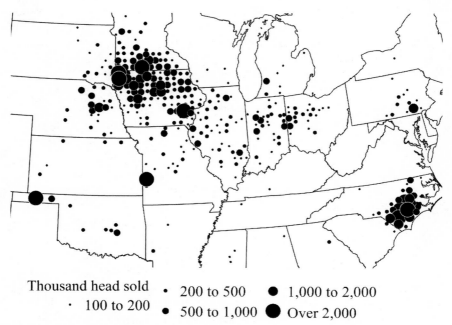

Thousand head sold   · 200 to 500   ● 1,000 to 2,000
· 100 to 200   • 500 to 1,000   ⬤ Over 2,000

Figure 6.4 Hog sales, 2012. *Source*: Created by authors, data from USDA Census of Agriculture 2012.

began processing more than one million hogs a year at the Guymon plant. Both Tyson Foods and Murphy Farms were attracted to hog production possibilities in western Oklahoma as well, although only the Seaboard venture remains as an integrated hog production and meatpacking operation.

Today's map of hog sales is virtually identical to one showing the number of hogs on contract (Figure 6.4). Northern Iowa and southern Minnesota are the heart of the industry and are also home to more than two dozen meatpacking plants that slaughter, dress, and ship the product. Like boxed beef, pork products leave the packing plant securely wrapped and ready for supermarket delivery. Eastern North Carolina's hog industry is more tightly concentrated, but the production system there operates in much the same fashion.

FOREIGN DIRECT INVESTMENT

What the map does not show is the corporate alignment of the meat packers who control the hog business. Having undergone bankruptcies and what seemed to be an endless chain of reorganizations, sales, and spinoffs, a period of stability emerged in pork packing when Smithfield Farms of Virginia purchased the Murphy operations in North Carolina in 2000. Smithfield, an old Virginia company, also acquired many of the other pork labels of long standing, including Armour, Eckrich, John Morrell, Farmland, and Patrick Cudahy. In 2013 Smithfield itself was purchased for $7.1 billion by Shuanghui International Holdings of Shanghai (Smithfield 2015). The sum represents the largest single Chinese investment to date in a U.S. company.

Foreign direct investment in the meat industry did not originate with the Shuanghui acquisition. In 2007 JBS S.A., a Brazilian company, purchased Swift & Co. from its then owner, Con-Agra of Omaha. JBS was founded in the 1950s by Jose Batista Sobrinho, a cattle rancher in Sao Paulo state who helped build the Brazilian beef industry into the world's largest. JBS owns more than two dozen meatpacking plants in Brazil and Argentina and has feedlots with a capacity of 200,000 head of cattle. JBS's acquisition of Swift substantially added to its North American holdings. JBS's acquisition also included the chain of twelve western U.S. and Canadian feedlots known as Five Rivers Ranch, which have a collective capacity of nearly one million cattle, and which had been started by Warren Monfort of Greeley, Colorado, one of the early "big beef" pioneers in the 1980s. JBS moved its American headquarters to Greeley and later acquired 75% control of Pilgrim's Pride, one of the largest U.S. poultry producers. In all, JBS is the world's largest food products company.

Brazilian beef, as prepared for sale on the world market by processors such as JBS, does not closely resemble the product which Swift & Co. made

famous. Brazilian cattle are sired by the *Bos indicus* (Indicine) breeds that
have long dominated in tropical meat industries. And more than four-fifths
of Brazilian cattle are kept on grass pastures (Ferraz and others 2009). The
cattle are mostly hybrids, produced by crossing the taurine and indicine lines,
the ancient distinction that separated cattle of European versus Indian origin.

To the surprise of some who presumed JBS would bring more of the Brazilian grassfed, Indicine type of production to the United States, the company
did the opposite. "Swift Black" is JBS's name for its superior, grain-fattened
beef, produced in Brazil for sale in the Brazilian market (JBS 2014). JBS has
also gone into the Australian beef industry—traditionally based on grass-
feeding—by constructing feedlots and meat-processing plants in Queensland
to produce grain-fed beef for the Australian market. Grainfed and grassfed
cattle production are both now worldwide and any attempt to claim that one
is preferable to the other would be even more difficult given the JBS grainfed
beef innovations.

REFERENCES

American Grassfed Association. 2015. *Directory.* http://www.americangrassfed.org/
about-us/.
Briggs, H. M., and D. M. Briggs. 1980. *Modern Breeds of Livestock.* New York:
Macmillan Publishing Co.
Broadway, Michael. 2007. Meatpacking and the Transformation of Rural Communi-
ties: A Comparison of Brooks, Alberta and Garden City, Kansas. *Rural Sociology.*
72(4): 560–582.
Davis, J. R., and H. S. Duncan. 1921. *History of the Poland-China Breed of Swine.*
Maryville MO: Poland-China History Association.
Dawson, H. C. 1913. *The Hog Book.* Chicago: Breeders Gazette.
Felius, M. 1985. *Genus* Bos. Rahway NJ: Merck & Co.
Ferraz, J. B. S., and P. E. de Felicio. 2009. Production Systems—An Example from
Brazil. *Meat Science* 84(2): 238–243.
Giuffra, E. J., M. H. Kijas, V. Amarger, O. Carlborg, J-T. Jeon, and L. Andersson.
2000. The Origin of the Domestic Pig: Independent Domestication and Subsequent
Introgression. *Genetics* 154: 1785–1792.
Harper, A. 2009. *Hog Production Contracts: The Grower-Integrator Relationship.* Vir-
ginia Cooperative Extension Service. [http://pubs.est.vt.edu/414/414-039/414-039.
html].
Hart, J. F. 2003. *The Changing Scale of American Agriculture.* Charlottesville: University of Virginia Press.
_____. 2007. Bovotopia. *Geographical Review* 97(4): 542–549.
Hart, J. F., and C. Mayda. 1997. Pork Palaces on the Panhandle. *Geographical Review*
87(3): 396–400.

Heath-Agnew, E. 1983. *A History of Hereford Cattle and Their Breeders*. London: Gerald Duckworth.

Henlein, P. C. 1959. *Cattle Kingdom in the Ohio Valley, 1783-1860*. Lexington: University of Kentucky Press.

Hudson, J. C. 1994. *Making the Corn Belt: A Geographical History of Middle-Western Agriculture*. Bloomington: Indiana University Press.

JBS. 2014. *JBS expands facilities to meet demand for Swift Black*. [http://www.jbs.com.br/en/media_center/press_releases/jbs-expands-facilities-meet-demand].

Krider, J. L., and W. E. Carroll. 1971. *Swine Production*. New York: McGraw-Hill.

McTavish, E. J., J. E. Decker, R. S. Schnabel, J. F. Taylor, and D. M. Hillis. 2013. New World Cattle Show Ancestry from Multiple Independent Domestication Events. *Proceedings of the National Academy of Sciences*. [www.pnas.org/cg/doi/10.1073/pnas.1303367110].

Rouse, J. E. 1977. *The Criollo: Spanish Cattle in the Americas*. Norman: University of Oklahoma Press.

Sanders, A. H. 1928. *A History of Aberdeen Angus Cattle*. Chicago: Lakeside Press.

Sauer, C. O. 1971. *Sixteenth Century North America*. Berkeley: University of California Press.

Smithfield. 2015. *The Smithfield Packing Company, Incorporated*. [http://www.vault.com/company-profiles/food-beverage/gwaltney-of-smithfield,-ltd-inc/company-overview.aspx].

Trow-Smith, R. 1957. *A History of British Livestock Husbandry to 1700*. London: Routledge and Kegan Paul.

U.S. Department of Agriculture, Agricultural Marketing Service. 2002. United States Standards for Livestock and Meat Marketing Claims. *Federal Register* 67 (79552, No. 250). Doc. No. LS-02-02.

_____. 2009. *Federal Register* 72(199): 58631–58637.

_____. 2016. Understanding AMS' Withdrawal of Two Voluntary Marketing Claim Standards. USDA Blog, January 20, 2016. [http://blogs.usda.gov/2016/01/20/understanding-ams-withdrawal-of-two-voluntary-marketing-claim-standards/].

Chapter 7

Poultry

The theme of abundance in the American food supply applies especially well to chickens, eggs, and turkeys. The United States currently produces 8.2 billion broiler (meat-type) chickens every year. Even after subtracting the roughly 18% share of that total which is exported to other countries, this leaves an average of 57 chickens for every American household to consume every year. The 1.82 billion dozen eggs produced annually translates to about 16 dozen eggs per household, and the nation's harvest of 195 million turkeys is quite a bit more than one per household. Although the total consumption of red meat in the United States is larger, chicken leads all other meats in popularity. And no country consumes more poultry and poultry products on a per capita basis than does the United States (USDA, Economic Research Service, 2015).

CHICKEN

Our common domestic chicken is descended from one or perhaps several species of jungle fowl (*Gallus gallus*) which today, as in ancient times, range in the wild from South India to China, Malaysia, and Philippines (Eriksson and others 2008). Domestications took place beginning about 7,000 years ago in India and later in scattered other areas within the species' range. Human interest in the chicken may have originated with the sport of cockfighting, but recognition of the nutritional value of the bird's eggs and meat surely led to its incorporation into human diets and gave the chicken a secure place in every civilization that followed (Pennsylvania State University 2015).

Westward diffusion of the domesticated chicken was both widespread and gradual. Chickens reached southeastern Europe roughly 5,000 years ago and were known in northwestern Europe within 2,000 years after that (Storey and others 2012). Chickens came to the Americas with the earliest explorers and soon became a part of human diet. Speculation has long been made as to whether an earlier, trans-Pacific contact took place (the elusive "pre-Columbian chicken") but the significant incorporation of chickens into North and South American cultures surely began with the contacts from Europe. While the turkey was a North American original which was taken to Europe almost immediately following its discovery, chickens originated in Asia and were brought to the Americas by Europeans.

Chickens are omnivores, easily fend for themselves, and, pound for pound, probably offer as much nutrition for humans as any animal that can be named. Chickens were found on practically every farm, from the English, French, and Spanish colonial days on to modern times. Egg production doubled in the United States over the first half of the twentieth century and chickens increased in number by 60% over the same period, first reaching a high of nearly one billion birds in 1943 (U.S. Department of Commerce, 1957).

THE BROILER INDUSTRY

The modern Broiler Belt, which extends across the entire breadth of the Southeastern states from Delaware to Texas, had its origin in developments that took place at opposite ends of that region during the 1920s. Histories of the broiler chicken industry generally credit Celia (Mrs. Wilmer) Steele, who had a commercial egg-laying business in Sussex County, Delaware, with having launched commercial broiler production in 1923 when she raised a flock of 500 chicks with the intention of selling them for meat (Johnson 1944). The business was so profitable that she built a 10,000-bird broiler house three years later and doubled her production again within another few years. Mrs. Steele's original broiler house (now located on the grounds of the Delaware Agricultural Experiment Station at Georgetown and listed on the National Register of Historic Places) illustrates the almost limitless opportunities in returns to scale that the early entrepreneurs saw in raising chickens for meat.

Two decades later Frank Perdue, who was similarly involved in a family poultry and egg business across the state line in Salisbury, Maryland, even more aggressively built a broiler chicken business that supplied markets on the East Coast. Perdue was a master of promotion and, through television advertising, he is credited with popularizing chicken among consumers nationally. Three more large poultry processing firms are located in Sussex County today, and together with Perdue (the second-largest broiler producer in the United States) they account for much of the Delmarva Peninsula's output of nearly a quarter of a billion meat-type chickens annually. In 2012

Figure 7.1   Heavy traffic in arriving chickens at a Delmarva poultry processing factory.
Source: Photograph courtesy of author.

Sussex County, Delaware, was the largest broiler producing-county in the United States by a wide margin (Figure 7.1).

A thousand miles west of Delaware similar developments were taking place in the northwest corner of Arkansas during the early decades of the twentieth century. Farmers in that region of rolling hills just outside the Ozarks were in need of an agricultural specialty that was more profitable than the vegetable and fruit farming they had been practicing (Riffel 2014). In 1931, John Tyson, who operated a trucking business based out of Springdale, Arkansas, was involved in transporting broiler chickens to markets in Kansas City and St. Louis. Tyson went into the broiler business himself and soon became one of Arkansas's leading producers. The family business, which was led by Tyson's son Don after the 1950s, grew into the giant Tyson Foods, which is now the second-largest agribusiness firm in the United States and the world's second-largest producer of beef, pork, and poultry.

Apart from the role of entrepreneurship, three trends guided these developments. One was the American consumer's steady increase in preference for chicken over other types of meat. A second trend was the favorable conditions for growth in the Southern states where farmers were in need of a replacement for their former reliance on cotton after the 1940s (Lord 1971). Southern farmers eagerly entered the new business of poultry raising which had been largely unknown to them in former times. In 1925 half the chickens in the United States were produced in the Midwest and per capita production nationally was about four birds per person. By the late 1960s most of the chicken was produced in the South and per capita consumption had nearly tripled. The third trend was a shift away from the traditional scale of general farming to an industrial scale of specialized production. This led to a blurring of lines between agricultural producers and the companies which controlled the infrastructure for marketing and processing. Today's system of contract growers supplying regionally dispersed processing plants grew out of all three trends.

A necessary precursor to these developments, however, was the recognition of broiler chickens as a category of farm produce. The full slogan, famously used by Herbert Hoover in his campaign for the U.S. presidency in 1928, was "a chicken in every pot and an automobile in every garage." Chicken was not considered a delicacy at that time, but it was not commonly found on the dinner table except perhaps on Sundays, and it was as often boiled, fricasseed, or stewed as it was fried. Young chickens—generally marketed as "spring chickens"—were more tender and were more suited to frying than were the older birds kept for egg production. The broiler was not necessarily a different breed of chicken, but its production was intended to be entirely for meat consumption. Costs of production were low since broiler chickens weren't kept on the farm long before they were sold.

Government-mandated meat rationing reduced the availability of both beef and pork to civilian consumers during World War II, but chicken was exempted from rationing and its production boomed. Having grown accustomed to more chicken in their diet, Americans kept consuming more chicken after wartime rationing ended. Per capita chicken consumption doubled from 1944 to 1964 and then doubled again by 1990. Chicken passed pork in per capita consumption in 1996 and it passed beef for the first time in 2010 (Bentley 2012). Health-conscious consumers are generally credited for this shift toward chicken, although the per capita consumption of chicken peaked at 60.9 pounds per person in 2006. The popularity of turkey has substituted for some of the demand for chicken in recent years.

Entrepreneurs like Frank Perdue, Don Tyson, and a dozen others took chances, expanded their production, and played a major role in boosting chicken consumption nationally. In the years after those two producers got their start, other entrepreneurs filled in the space between Delaware and Arkansas with operations that closely followed the lead of the early companies (Figure 7.2). By 1969 nearly 90% of the national supply of meat-type chickens was raised in the South, even though egg production and other types of poultry raising remained more geographically dispersed.

Gold Kist poultry, based in Carrollton, Georgia, west of Atlanta, began as a cotton marketing cooperative organized by local farmers in the 1930s.

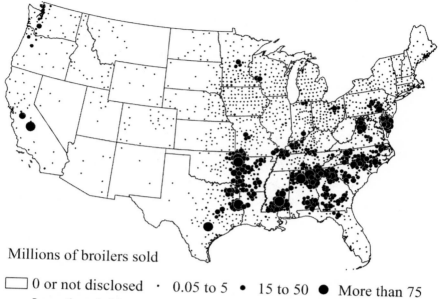

Millions of broilers sold

☐ 0 or not disclosed	· 0.05 to 5	• 15 to 50 ● More than 75
· Less than 0.05	• 5 to 15	• 50 to 75

Figure 7.2 Broilers sold, 2012. *Source*: Created by authors, data from USDA Census of Agriculture 2012.

Gold Kist diversified into broiler production in the 1940s and by the 1970s had become a vertically integrated poultry and feed conglomerate with eight regional broiler complexes each with its own network of growers, hatcheries, and processing plants (Hart 1980). Gold Kist was purchased by a Texas poultry company, Pilgrim's Pride, in 2006. Pilgrim's Pride was founded in 1946 by Aubrey Pilgrim of Pittsburg, Texas, who operated a feed business to supply nearby farmers who raised chicks. Sanderson Farms of Laurel, Mississippi, similarly originated as a locally owned chicken-feed supply business in the late 1940s and grew to become one of the largest corporations in the state of Mississippi.

Almost from the start, the broiler business was organized along lines that set it apart from other types of agricultural activities (Perry and others 1999). Feed-producing operations played a large role in the broiler business compared with beef and pork because broilers were a Southern specialty and the South lacked sufficient quantities of feed grains. In the Midwest hogs and cattle were fed from corn crops grown on the very farms where they were kept. But the Southern poultry producer had to purchase quantities of corn and soybeans that were imported from states farther north and then processed at a local feed mill. Some feed companies also acted as financial institutions, advancing credit to the farmers who then purchased chicks from local hatcheries. Growers, feed mills, hatcheries, and poultry processors were linked businesses but they remained under separate control and ownership until the larger poultry companies began to vertically integrate all of the components under their corporate control (National Chicken Council 2014).

Those who raised chicks to market-weight broilers were vulnerable to price fluctuations. Some growers had lost their farms to lending institutions when poultry prices dropped, and they were unable to repay the loans for chicks and feed that kept their operation in business. Contract growing was introduced in the 1960s as a way to shield farmers from some of those risks. Under this system farmers sign an agreement with a poultry processor to feed a batch of chicks in a specified time using a diet stipulated by the company. Day-old chicks are picked up at a company-owned hatchery and delivered directly to the grower's broiler houses. Feed, medicines, and other necessities are delivered as well. The birds are fed for seven to ten weeks, depending on the size desired. They are then picked up by the processor company, and within a matter of hours are inside the processing plant being prepared for sale. Chicken farmers effectively became employees of the poultry processors as a result of this system, but they stand fewer financial risks and are better able to plan their business.

Soybeans, which are an important source of chicken feed, are raised in the South, but the very restraints imposed by poor land and hilly topography that led Southern farmers into the poultry business in the first place also meant that their farms were not well suited for crop production. They did not have to look far to find an alternative source. Feed grains are grown in abundance in the Mississippi, Illinois, and Ohio River Valleys where farmers are used to selling their crops for cash. Once harvested, corn and soybeans are trucked to riverbank terminals, barged first downriver and then up the Tennessee River to ports such as Decatur and Guntersville, Alabama. Large poultry feed mills, built next to the barge tie-ups on the banks of the Tennessee River, produce the chicken feed which is moved by truck or rail to the poultry-raising areas of Alabama and North Georgia. Broiler industries elsewhere across the South, from Delaware to Oklahoma and Texas, are similarly supplied with Corn Belt–raised grain crops brought in trainload quantities to the poultry production areas. Feed grains from the Midwest thus form an important part of the chickens' diet.

Live chickens are transported relatively short distances to the packing plants (Figure 7.3). The major poultry companies dominate activities within regions that are about half the size of the average state. A network of growers is found within short-distance access of every processing facility within a production region. Smaller companies are scattered among plants owned by the large producer such as Tyson and Pilgrim's Pride. All of the packing

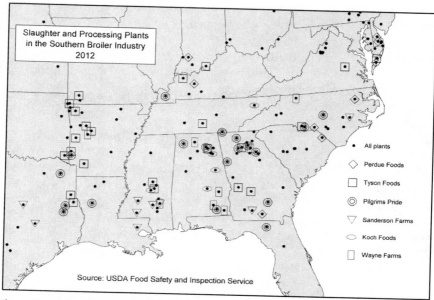

Figure 7.3 Slaughter and processing plants in the southern poultry industry. *Source:* Created by authors, data from USDA Food Safety and Inspection Service.

plants either own or have access to extensive cold storage facilities within their regions of production. From there the birds are transported to market by refrigerated trucks.

EGGS

Poultry breeding is an ancient preoccupation, which explains, in part, the number and abundance of chicken breeds around the world having contrasting appearances (Akers and others 2002). Throughout most of history the objective of breeding was to produce a superior animal that would attract attention at poultry shows. Color of plumage, shape of the comb, and the general appearance of the bird were the focus of breeder's efforts. Wyandottes, Plymouth Rocks, Rhode Island Reds, and Leghorns were some of the American chicken breeds that were subject to refinement for appearance's sake. Breeds from Asia and Europe drew attention at poultry shows as well.

Breeding for appearance brought recognition to individual poultry breeders, but the results were known only to a small circle of like-minded individuals and the monetary returns were small. The introduction of genetic science into agriculture during the early decades of the twentieth century gradually

changed the objective of breeding to focus more on characteristics of production and less on outward appearance (Mississippi State University 2014).
Genetic principles revolutionized corn breeding, for example, and they had
a major impact on chicken breeding as well. The recognition of heterosis
(hybrid vigor) stimulated these efforts.

Selection of chickens for disease resistance, size, rapid weight gain, onset
of egg production, and quality of meat soon led to a divergence between
meat-type chickens and egg-laying chickens (University of Georgia 2012).
The breed of the chicken, such as Rhode Island Red or Plymouth Rock, mat-
tered less than the traits favorable to meat production versus egg production.
Breeding a laying hen that would produce eggs at a younger age and over
a longer span of time was one objective. At present, laying hens begin producing eggs at the age of 18 weeks and produce about 200 eggs in their first
year. Nationally in the United States, some 339 million birds lay 92 billion
eggs per year, an average of 271 eggs per hen. Egg and broiler production
require different types of care for the birds, including different styles of
housing to match life-cycle requirements. Chickens are day-length-sensitive
when it comes to their egg-laying behavior, an issue that does not impact
broiler production (University of Georgia 2012).

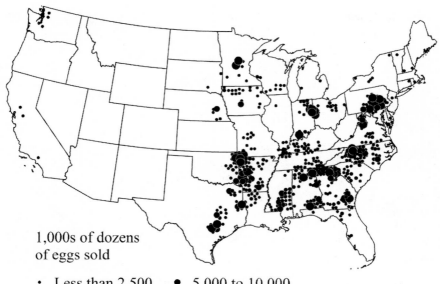

1,000s of dozens
of eggs sold

· Less than 2,500   ● 5,000 to 10,000
· 2,500 to 5,000   ● More than 10,000

Figure 7.4   Eggs sold, 2012. *Source*: Created by authors, data from USDA Census of
Agriculture 2012.

The United States produced more than a billion eggs for the first time in the late 1880s and by 1929 the total had reached 2.9 billion. The top egg-producing states in the first half of the twentieth century were Iowa, Missouri, Illinois, Ohio, and Indiana—the heart of the Corn Belt. Because eggs are both fragile and perishable, production near the market place has always been desirable. As population moved westward, so did egg production and by 1954 California was the largest egg-producing state. Pennsylvania, New York, and New England also were major egg producers.

Despite these differences, eggs and broilers are produced in roughly the same kinds of settings today as they were in the past (Figure 7.4). In 2012 the top six broiler-producing states were included among the top nine egg producers: Georgia, Alabama, Arkansas, North Carolina, Mississippi, and Texas. Pennsylvania is the largest egg producer, however, and Ohio, Indiana, and Iowa are included in the top ten egg states. Egg production thus retains its long-established connection with the Corn Belt, even though much of the industry is now located in the Southern states, along with the broilers. California and New York are no longer major egg producers.

TURKEYS

The domestic turkey, which was bred from the native North American wild turkey, *Meleagris gallopavo,* has undergone several transformations in both size and shape since it became a commercial product (Smith 2006). The name, "turkey," is a misnomer resulting from an incorrect identification of the bird's origins made by early European-Americans who thought the bird was a Guinea fowl of the sort raised in Turkey. The wild turkey is a large bird, at home in the fringes of hardwood forests. It was (and still is) a favorite of hunters in the Eastern Woodlands.

The first turkeys kept on American farms were similarly large, brown-feathered birds, essentially the same as those found in the wild. Roughly 10% to 15% of American farms raised at least a few turkeys back into the late nineteenth century, and there was an annual harvest of several million farm-raised birds at that time. The public's fondness for turkey stimulated attempts at breeding to increase the bird's market value. During the Great Depression decade (1929–1939) turkey production increased by 65%, even though the number of farms producing turkeys decreased by 50% at the same time. The trend toward larger production but on fewer farms in the 1930s signaled the emergence of turkey raising as yet another specialty farm business.

The bird itself had some traits that commercial growers wanted to change. In the 1930s turkeys were large (18–25 pounds) and had a narrow breast without much meat. The ratio of dark meat to white meat was the reverse of what consumers preferred and the dressed birds were too large to fit into the home refrigerators then available. Working from the genetic stock of four different turkey breeds, poultry breeders at the USDA's Beltsville Research Station outside Washington, DC, produced a "new" turkey that changed some of these traits. The Beltsville Small White turkey made its debut in 1947 (USDA Agricultural Research Service 2015). It was almost immediately adopted by turkey breeders across the nation and became the genetic foundation for practically all turkeys produced since that time, although its dominant position in the marketplace did not last for long.

The demand for a turkey that was larger in size than the Small White and had more breast meat produced yet another breeding revolution, resulting in the Broad Breasted White, which became the industry standard by the mid-1960s. Government scientists and commercial breeders had reduced the turkey's size and gotten rid of the brown feathers for white. They then reduced the proportion of dark meat and increased the bird's size once again, creating an animal that was largely breast meat. Broad Breasted Whites require artificial insemination and cannot reproduce sexually because of their large breast size. About 285 million turkeys of this type are sold in a typical year at present.

As consumer demand for turkey grew, the production per farm increased and the number of turkey-producing farms decreased further. Yet another revolution in consumer demand took place when turkey became a year-around food item and not just the centerpiece of the annual Thanksgiving dinner. Between the early 1980s and 1990s turkey production increased from 2.5 million pounds per year to more than 6 million pounds.

At the national scale today, the map of turkey production resembles that of egg production, but with a marked concentration in Minnesota, Iowa, and South Dakota (Figure 7.5). In 1940 Earl Olson founded the Jennie-O turkey company of Willmar, Minnesota, which was purchased by Hormel in 1986. Wallace Jerome of Barron, Wisconsin, founded a parallel turkey raising, processing, and marketing company known as the Turkey Store which also was acquired by Hormel, in 2001. Jennie-O Turkey Store is the second-largest turkey processor at present, slightly smaller than Butterball LLC, a North Carolina–based company jointly owned by Seaboard Farms and Maxwell Farms LLC.

Like the broiler chicken industry, genetics and hatcheries in the turkey business are under the control of major companies. Life Science Innovations, incorporated as a privately held technology park in central Minnesota, coordinates research and egg production with an annual hatch of 45 million poults which are delivered to individual growers in the Middle West. North Carolina has a similar concentration of genetics and hatcheries businesses. Minnesota and North Carolina are the two leading turkey-producing states (National Turkey Federation 2015).

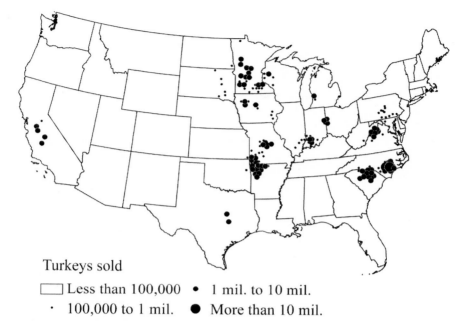

Turkeys sold

☐ Less than 100,000   • 1 mil. to 10 mil.
· 100,000 to 1 mil.   ● More than 10 mil.

Figure 7.5   Turkeys sold, 2012. *Source:* Created by authors, data from USDA Census of Agriculture 2012.

THE AVIAN INFLUENZA OUTBREAK OF 2015

The growth of food production over time is not always a success story. Setbacks do occur and they call forth a need for reappraisal of past practices. In December 2014 several cases of Avian Influenza (AI) were recorded in mixed flocks of domestic poultry on the West Coast. By March, 2015, the virus had been detected in a flock of turkeys in Minnesota and soon thereafter widely separated flocks of turkeys in Missouri and Arkansas were affected. Joint efforts of the USDA's Animal and Plant Health Inspection Service (APHIS) and the Centers for Disease Control included destroying all birds in every flock where AI had been confirmed (USDA, Animal and Plant Health Inspection Service, June 4, 2015).

The AI virus raged from March through June of 2015, affecting some 223 flocks of turkeys and laying chickens. In all, more than 48 million birds were euthanized to prevent further spread. The Minnesota turkey industry was hard hit, especially in its core production area around Willmar. Chicken egg-laying operations—which typically house more than ten times as many birds as turkey farms have—were devastated in northwest Iowa and eastern Nebraska.

Wild ducks and geese spread the virus but they are typically not affected by it. Evidence that wild birds are the long-distance transmitters includes the association of outbreaks with the transcontinental flyways of migrating birds, especially waterfowl. Sudden jumps of several hundred miles in various directions spread AI from Minnesota into South Dakota, then North Dakota, and from Minnesota into Iowa and Nebraska. APHIS scientists later confirmed that wild birds were responsible for introducing AI into commercial poultry (USDA, Animal and Plant Health Inspection Service, July 15, 2015).

AI also spread rapidly at the local scale once it was first introduced. The virus spreads mainly when healthy birds come in close contact with affected birds. The sheer enormity of disposing of millions of euthanized birds required the efforts of hundreds of government workers whose own mobility probably caused the virus to spread at the local scale. Other factors causing AI to spread included sharing of equipment and the mobility of personnel between infected and uninfected farms. The presence of strong winds also appears to have affected the spread of the disease at the very local scale.

Outbreaks of highly pathogenic AI are likely to occur again. Keeping chickens and turkeys indoors where they cannot come in contact with wild birds would seem to be a sound practice, although the "cage-free" or "free-range" label has become a favorite of poultry marketers and is virtually mandated by federal Certified Organic standards. Since only a single case of AI is sufficient for an entire flock to be destroyed, the number of birds per flock is perhaps not as significant as is the geographical clustering of poultry-raising operations. Once introduced into a locality, the virus can spread rapidly in all directions. Several geographical scales of disease spread are thus involved.

The world's annual production of chickens—some 40 billion birds per year, or more than five times the world's human population—has remained fairly constant over the past decade after years of steady increase. Epidemics such as AI might eventually hold total numbers below a certain level, but there is no evidence that this is happening even in the face of massive reductions like the AI episode of 2015. Market demand, rising levels of living, the need for protein in the human diet, and changing consumer tastes have a much greater influence on the total level of production. As in other food sectors, there seems to be no logical upper limit to how much poultry the world can consume.

REFERENCES

Akers, D., P. Akers, and M. A. Latour. 2002. *Choosing a Chicken Breed: Eggs, Meat, or Exhibition*. Purdue University Cooperative Extension Service [https://www.extension.purdue.edu/extmedia/as/as-518.pdf].

Bentley, J. 2012. *U.S. Per Capita Availability of Chicken Surpasses That of Beef.* USDA, Economic Research Service. [http://www.ers.usda.gov/amber-waves/2012-september/us-consumption-of-chicken.aspx#.VbeX3flVhBc.]

Eriksson, J., G. Larson, U. Gunnarrsson, B. Bed'hom, M. Tixier-Boichard, L. Strom-stedt, D. Wright, A. Jungerius, A. Vereijken, E. Randi, P. Jensen, and L. Andersson. 2008. Identification of the Yellow Skin Gene Reveals a Hybrid Origin of the Domestic Chicken. *PLOS Genetics* 4(2). [http://journals.plos.org/plosgenetics/article?id=10.1371/journal.pgen.1000010].

Hart, J. F. 1980. Land Use Change in a Piedmont County. *Annals, Association of American Geographers* 70(4): 492–527.

Johnson, H. A. 1944. *The Broiler Industry in Delaware.* University of Delaware, Agricultural Experiment Station, Bulletin No. 150.

Lord, J. D. 1971. The Growth and Localization of the United States Broiler Chicken Industry. *Southeastern Geographer* 11(1): 29-42.

Mississippi State University. 2014. *History of the Mississippi Poultry Industry.* [http://msucares.com/poultry/commercial/history.html].

National Chicken Council. 2014. *U.S. Chicken Industry History.* [http://www.nationalchickencouncil.org/about-the-industry/history/].

National Turkey Federation. 2015. [http://www.eatturkey.com/].

Pennsylvania State University, College of Agricultural Sciences, Penn State Extension. 2015. *History of the Chicken.* [http://extension.psu.edu/animals/poultry/topics/general-educational-material/the-chicken/history-of-the-chicken].

Perry, J., D. E. Banker, and R. Green. 1999. *Broiler Farms' Organization, Management, and Performance.* USDA, Economic Research Service, Agricultural Information Bulletin No. AIB-748.

Riffel, B. E. 2014. Arkansas Poultry Industry. *The Encyclopedia of Arkansas History and Culture.* [http://www.encyclopediaofarkansas.net/encyclopedia/entry-detail.aspx?entryID=2102].

Smith, A. F. 2006. *The Turkey. An American Story.* Urbana: University of Illinois Press.

Storey, A. A., J. S. Athens, D. Bryant, M. Carson, K. Emery, S. deFrance, C. Higham, L. Huynen, M. Intgoh, S. Jones, P. V. Kirch, T. Ladefoged, and P. McCoy. 2012. Investigating the Global Dispersal of Chickens in Prehistory Using Ancient Mitochondrial DNA Signatures. *PLOS Genetics.* [http://journals.plos.org/plosone/article?id=10.1371/journal.pone.0039171].

USDA, Animal and Plant Health Inspection Service. 2014. *Update on Avian Influenza Findings*, June 4, 2015. [http://www.aphis.usda.gov/wps/portal/aphis/ourfocus/animalhealth/sa_animal_disease_information/sa_avian_health/ct_avian_influenza_disease].

_____. 2015. Stakeholder Announcement. *APHIS Releases Partial Epidemiology Report on Highly Pathogenic Avian Influenza.* [http://www.aphis.usda.gov/animal_health/animal_dis_spec/poultry/downloads/Epidemiologic-Analysis-July-15-2015.pdf].

USDA, Agricultural Research Service. 2015. *A Brief History of Turkey Research and the Role of the Beltsville Agricultural Research Center.* [http://www.ars.usda.gov/sp2UserFiles/Place/80000000/Partnering/TurkeySuccess.pdf].

USDA, Economic Research Service. 2015. *Poultry and Eggs, Production and Trade.* [http://www.ers.usda.gov/topics/animal-products/poultry-eggs/statistics-information.aspx].

University of Georgia, College of Agricultural and Environmental Sciences, Cooperative Extension. 2012. Seven Reasons Why Chickens are NOT Fed Hormones. *Poultry Housing Tips* 24(4): 1–2. [https://www.poultryventilation.com/tips/vol24/n4].

U.S. Department of Commerce. 1957. *Historical Statistics of the United States to 1957.* Washington DC, 1961.

Chapter 8

Fruits and Vegetables

Production of the thirty largest fruit and nut crops and the thirty largest vegetable crops requires 7.1 million acres of land in the United States today. The acreage, which is divided about equally between fruits and nuts versus vegetables, accounts for less than 3% of the nation's total cropland. But these are the supermarket crops, the high-value perishable items, whose value per acre exceeds all others. Fruits and vegetables yield more income and require more labor and capital per acre compared with other crops, and they typically are more sensitive to environmental factors during the growing period and at harvest. They require more attention but repay the extra effort through their greater cash returns.

These crops are by no means evenly spread across the country. The state of California is by far the largest producer, accounting for over 30% of the vegetable acres and nearly 60% of fruit and nut acreage. Production is concentrated even within California where just four counties (Fresno, Kern, Tulare, and Merced) contain nearly one-fourth of all fruit- and nut-producing acres in the United States. On the other hand, eleven of California's fifty-eight counties report no acreage in fruit and nut crops at all. The intense concentration of production at both the state and county scales is the result of several factors, among which climatic variables such as temperature and moisture are of great significance.

IRRIGATION AND CLIMATE

California uses more than 23 million acre-feet of irrigation water every year, the most of any state and nearly three times as much as Nebraska, which is the second-largest agricultural water user. California's water consumption

97

amounts to more than one-fourth of all the water used for irrigation in the United States. California's year-round growing season is aided by warm to hot temperatures all year which are excellent for producing the biomass that all crop plants consist of. But without the addition of irrigation water the state's fruit and vegetable crops would be next to impossible (Figure 8.1). In most California counties more than 95% of the farms are irrigated and those that are not irrigated typically are cattle-grazing enterprises or others that do not focus on crops.

California is not the only state where fruit and vegetable crops depend heavily on irrigation. Washington's apples, Colorado's peaches, New Mexico's chili peppers, Idaho's potatoes, and Texas's grapefruit are equally tied to a regular supply of irrigation water. In fact nearly all crops grown in the western states require more water than rainfall alone can provide. Even Wisconsin's and New Jersey's vegetable crops, grown on sandy soils that need supplemental water, often are irrigated despite the relatively moist climates found in those two states. Irrigation water pumped up from underground aquifers or diverted from streams is applied to fruit and vegetable crops from coast to coast. Water supplied from the ground up, rather than downward from the atmosphere, has the added benefit of keeping the fruits and leaves of plants dry which also makes them less susceptible to disease and insect damage.

Temperature and moisture are the two crucial environmental factors affecting crop production. Of these two, the supply of moisture is obviously the

Figure 8.1   Irrigated orchards in California's San Joaquin Valley. *Source*: Photograph courtesy of author.

easier to supplement. Plant growth responds rapidly to increased warmth in the growing season. For most crops the warmer the temperature, the faster the growth. Plants growing under hotter conditions require more water to maintain their existence, whereas cool temperatures often render some water as surplus because plants cannot make use of it. For these reasons, many varieties of fruit and vegetable production have migrated toward the hotter and the drier environments over the past century and more.

FRUIT AND NUT CROPS

The national map of fruit and tree nut acreage combines the summer-only crops of the northern states with the nearly year-round production offered in the warm-winter climates of Florida and California (Figure 8.2). California and Florida account for three-fourths of the total value of U.S. fruit and nut production. Washington, Oregon, and Georgia—not as warm but with long growing seasons—produce another 15% of the value.

Grapes are the most widely grown fruit crop, covering more than one million acres of vineyard land, 85% of which is in California. The state's grape

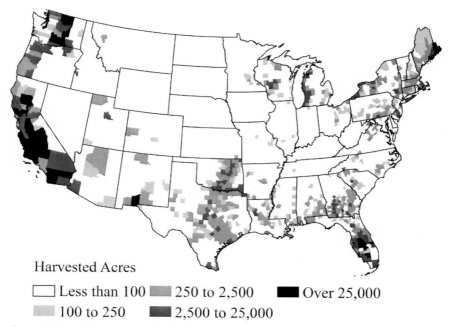

Figure 8.2   Fruit and tree nut crops, 2012. *Source*: Created by authors, data from USDA Census of Agriculture 2012.

acreage is divided in terms of the use made of the fruit: wine (63%), raisins (25%), and table use (12%) (California Department of Food and Agriculture 2015). Wine grape production used to be concentrated in the Coast Range valley north and south of San Francisco where the wine industry was firmly established. Table and raisin grapes were the crops grown in the southern San Joaquin Valley. Today, however, wine grapes are grown widely in the state and the wine industry has expanded into the San Joaquin Valley. Hot growing-season temperatures, abundant sunshine, and copious amounts of irrigation water support the grape crops.

Washington is the second-largest grape producing state, and although the Pacific coastal mountains of Washington are some of the wettest places in the nation, that is not where grapes are grown. Washington's grapes, like California's, are grown at low elevations on the rain shadow side of the mountains where maximum growing-season temperatures are achieved under cloudless skies and where irrigation water channeled down from snow-fed reservoirs supports the crop's growth.

Hot temperatures are not necessary to produce all varieties of grapes. New York ranks third among the states in grape acreage, some of which is used for growing champagne varieties. In a belt of counties along Lake Erie, Concord grapes are grown for manufacture into sweet grape juice and jam. More than 6,000 acres of vineyards in southwestern Michigan produce Concord grapes for manufacture into grape juice. Despite the large domestic production, more than half of the grapes consumed in the United States are imported. Grapes grown in Chile are the mainstay of U.S. table grape sales during the winter months when little domestic production is available (Huang and Huang 2007).

Following grape acreage in rank at the national scale are almonds (737,000 acres), oranges (508,000 acres) and apples (294,000 acres). Almonds, which are consumed largely as snack food, are grown almost exclusively in the Sacramento and San Joaquin Valleys of California. Orange tree acreage is divided between Florida (two-thirds) and California (one-third). Nearly two dozen counties in those two states have more than 1,000 acres of oranges each. Ninety percent of Florida's orange harvest is used to produce orange juice but citrus production in the state is threatened by disease problems (Florida Citrus Mutual 2015). Approximately 40% of U.S. orange juice consumption is based on imports, much of it from Brazil. South Africa, Mexico, Chile, and Australia all export oranges to the United States to supply the fresh fruit market.

Florida's east coast also produces two-thirds of the grapefruit grown in the United States, with much of the rest coming from the Rio Grande Valley of southernmost Texas. Grapefruit has declined in popularity in recent years, in part because of health warnings that the fruit intensified the effects of certain medications. Nationwide grapefruit production has declined more than 50% since 2000 (Perez and Plattner 2015).

Florida once had a thriving Key lime industry until much of it was destroyed in a hurricane in 1926. Growers then switched to the larger, greener, juicier Persian limes, which were produced in South Florida for many years thereafter. Along with a few citrus orchards in California, these crops supplied the U.S. market for limes, a fruit which is widely consumed, although not in large quantities.

A fruit-tree disease known as citrus canker has reduced Florida's lime crop to very low levels during the past decade. Although part of Florida's acreage remains, as does California's, nearly all the limes purchased in the United States today are imported from Mexico, the world's largest lime producer. Meanwhile, U.S. per capita consumption of limes has increased from less than one pound per person in 1990 to more than 2.5 pounds today. Mexico's large crop of limes has come about largely for the purpose of selling to the U.S. market (Plattner 2014).

The geography of apple production reveals a variety of historical and environmental influences. The state of Washington produces about half of the nation's apple crop, but only two states (Alaska and Hawaii) reported no commercial apple production at all in 2012. Apples are one of the most widely grown crops both because of their popularity and because of the variety of conditions under which they will grow. Milder winter temperatures around Lakes Michigan, Erie, and Ontario permit apple production in the Great Lakes states (Michigan Apple Committee 2015). Steep slopes, favoring cold air drainage, allow apple trees to be winter hardy in Pennsylvania and New England.

Counting the minimum number of counties required to total half the acreage in a crop is a measure of concentration in its pattern of production. Compared with most other crops, apples are relatively dispersed. Half the productive apple-orchard acres are in seven counties, located in Washington, New York, Pennsylvania, and Michigan. Most fruit and nut crops are more concentrated than that. Fifteen of the thirty most-produced crops are concentrated in one or two counties (Table 8.1).

Some crops, such as dates (which thrive in the dry heat of the Sonoran desert), wild blueberries (which abound in Eastern Maine's cold, acidic soils), and cranberries (raised in the sandy, acid-marsh soils of central Wisconsin and Cape Cod), are grown in only a few places because of environmental reasons. Other crops, such as raspberries, are concentrated in small areas almost by chance. Whatcom County, Washington, north of Seattle, has the largest raspberry acreage in the United States, although raspberry canes

Table 8.1   Concentration of fruit and nut crops, 2012

Crop	Half the crop acreage is in one or two counties in:
Apricots	California
Avocadoes	California
Blueberries, wild	Maine
Dates	California
Guava	Florida
Grapefruit	Florida
Lemons	California, Arizona
Limes	California, Arizona
Mangoes	Florida
Papaya	Hawaii
Pineapples	Hawaii
Pistachios	California
Raspberries	Washington
Strawberries	California
Tangerines	California

Crop	Half the crop acreage is in three or four counties in:
Almonds	California
Blackberries	Oregon
Cherries, tart	Michigan
Cranberries	Wisconsin, Massachusetts
Hazelnuts	Oregon
Oranges	Florida, California
Pears	Oregon, Washington
Plums and Prunes	California
Walnuts	California

Crop	Half the crop acreage is in five to eight counties in:
Apples	Washington, New York, Pennsylvania, Michigan
Blueberries, tame	Michigan, New Jersey, North Carolina, Washington, Georgia, Oregon
Cherries, sweet	California, Oregon, Washington
Grapes	California
Peaches	California, South Carolina, Georgia

Crop	Half the crop acreage is in thirty-four counties in:
Pecans	New Mexico, Georgia, Texas, Oklahoma, Louisiana

Source: USDA Census of Agriculture

cover only a small fraction of this county which is located just south of the Canadian border. California has fewer acres in raspberries, but it has higher yields per acre, and it is the leading state for fresh raspberry production. But even Washington's and California's raspberry crops are insufficient to meet American demand. Raspberry imports from Mexico and Chile, especially during the winter season, are equivalent to about 18% of the U.S. domestic crop. Nearly all raspberry imports are marketed fresh (Huang and Huang 2007; Geisler 2012b).

Four counties in Oregon's Willamette Valley produce most of the blackberries harvested in the United States. Of Oregon's total 2014 production of 44.86 million pounds, 92% was used for processing into juice, jam, and frozen concentrates. Fresh blackberries would be almost unavailable in the United States were it not for imports. In the same year, the United States imported 95.7 million pounds of fresh blackberries mainly from Mexico. As with raspberries, the domestic blackberry crop is processed while the imports are sold fresh (Geisler 2012b).

At present, nearly half of the fresh fruit, two-fifths of the canned fruit, and one-third of the fruit juice consumed in the United States are imported from other countries (USDA Economic Research Service 2015; Ferrier 2014). Mexico is the largest supplier, followed by Chile (a variety of fresh fruit), Brazil (orange juice) and China (apple juice). Excluding bananas, fresh fruit imports have grown from roughly 12% of domestic consumption in 1990 to around 25% today. The increased presence of fresh fruit in the American diet clearly owes much to imports. The NAFTA agreement has eliminated most tariffs on agricultural imports from Mexico as well as from Canada (Zahniser and others 2015). Chile, with its counter-seasonal Southern Hemisphere production schedule, does not compete directly with American producers in the fresh fruit market and has low tariffs on its products entering the United States.

Pecan production contrasts sharply with the foregoing observations about small-area concentrations of a crop. Some 25,000 acres of irrigated pecans growing in Doña Ana County, New Mexico, is the largest single source of pecans. But one has to include the acreage in another 33 counties in five states to get as much as half of the total harvested area of pecans. Pecans are produced commercially in all states where the tree is winter-hardy and they are the main source of cash income for many small farms in a belt from southern Missouri across Oklahoma into Texas. Georgia is the largest producer with typically about 30% more pecans than New Mexico. But Texas, Arizona, Oklahoma, and Alabama also produce pecans in quantity.

The pecan, *Carya illinoinensis*, is a North American native, which may be a key to its wide adaptation to mild-winter environments in the American South as well as in Mexico and Central America. The U.S. imports pecans from Mexico and exports pecans to Hong Kong and Southeast Asia in roughly equal amounts. Per capita pecan consumption has varied only slightly for many years around an average of 0.5 pounds per person (Perez and Plattner 2015). Pecans illustrate that specialty crop production is not necessarily a business involving small numbers of farms in a limited area.

VEGETABLES

The 3.25 million acres of vegetables raised in the United States every year are somewhat more dispersed than the fruit acres (Figure 8.3). To get half the vegetable acres requires two dozen counties ranging over eight states. The greater dispersion of vegetables is due, in part, to the greater climatic variability permissible with crops such as potatoes and sweet corn, which are two of the largest crops in terms of national acreage. Vegetable crops are roughly sorted by seasonal temperature differences into two concentrations: a northern region of summer-only production, and a coastal/subtropical region of year-round production.

Crops like artichokes, broccoli, Brussels sprouts, and cauliflower have been grown around the northern shores of the Mediterranean Sea ever since Roman times. Not surprisingly, they are easily cultured in California, which has the only sizable area of Mediterranean climate in the United States. Garlic, carrots, and spinach originated in Ancient Persia and they, too, are well adapted to a climate like California's. Celery, lettuce, and asparagus were likewise known in Ancient Greece and Egypt. All of these crops are suited

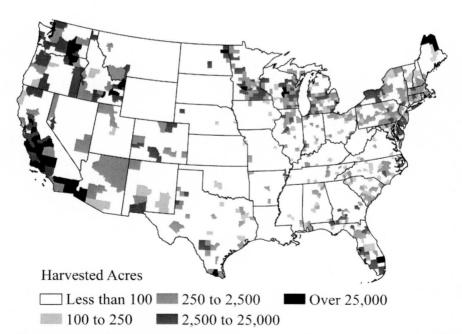

Harvested Acres

☐ Less than 100   ▨ 250 to 2,500   ■ Over 25,000
▨ 100 to 250   ▨ 2,500 to 25,000

Figure 8.3   Vegetable crop acreage, 2012. *Source*: Created by authors, data from USDA Census of Agriculture 2012.

to California's long growing season, and need irrigation during the hot, dry summer. California produces the bulk of the domestic crop in all of these cases.

The fact that California's climate is well suited to many crops of Mediterranean origin does not mean that other parts of the world cannot compete on the same basis. Garlic is an example. China began exporting garlic to the United States in the 1990s until its share of the U.S. garlic market was so large that California garlic farmers sought federal protection. In 1994 the Commerce Department placed a 376% tariff on Chinese garlic imports claiming that China was dumping garlic on the U.S. market. Imports shriveled immediately and U.S. garlic acreage reached 42,000 acres by 1999. But China is a formidable competitor. Its several climates permit year-round garlic production and its agricultural production costs are low. Chinese garlic exports to the United States began to grow once again, despite the tariff, as some California garlic growers switched to other crops (Huang and Huang 2007). By 2012 the U.S. garlic crop had fallen back to little more than 20,000 acres.

One of the largest vegetable acreages in the United States is the lettuce crop of California's Salinas Valley to which some 135,000 acres were devoted in 2012, all of them in one county (Monterey). Salinas Valley lettuce is supplemented by 70,000 acres grown in Yuma County, Arizona, using water diverted from the Colorado River; and another 42,000 acres in California's Imperial Valley, which also uses Colorado River water. Most of the U.S. spinach crop comes from those same areas. Because the production of lettuce and spinach is possible year-round in those three areas, U.S. imports of the two crops are small.

South Florida is the largest year-round producer of fresh vegetables in the United States. The crops in which it specializes, such as tomatoes, sweet corn, cucumbers, bell peppers, and snap beans, are among the crops most often imported from other countries in the winter season when only Florida and far southern Texas can provide the U.S. supply. But those same crops are also grown as summer vegetables farther north.

New Jersey's "Garden State" counties—in the southern portion of the state, between the Atlantic Ocean and Delaware Bay—produce many of the same crops grown in South Florida, but the New Jersey contribution comes during the summer season only. Some crops move northward with the season, from South Florida (winter), through north Florida, Georgia, South Carolina, North Carolina, Delaware, New Jersey, and, finally, western New York. Sweet corn, cabbage, potatoes, tomatoes, eggplant, cucumbers, greens (kale, mustard, collard), and watermelons are all crops that follow such a northward-moving cycle over the year. Much of the produce is destined for urban markets in the northeastern states.

AMERICAN-ORIGIN CROPS

The vegetable crops raised just in California differ from those produced else-where in the United States on the basis of climate, but also because of the provenance of the crops. Crop plants that are native to the Americas have a more widespread pattern of production (Table 8.2). Potatoes, which are the most widely grown American vegetable crop, are native to Andean South America, as are tomatoes. Corn growing is traceable to southern Mexico. The *Cucurbits* pumpkin and squash are of Meso-American origin, as are common beans (*Phaseolus vulgaris*), including green (snap or string) beans and dry edible beans. All of these plants are adapted to diverse environments in North America and, indeed, were found growing here in abundance by the early European travelers who described what they saw in the land.

Potatoes are one of the world's most important food crops, ranking fourth in total production following maize, wheat, and rice. After the Spanish dis-covered potato in South America and took it back to Europe, efforts were made to improve the quantity and quality of the crop. Centuries later a major advance was made when agricultural pioneer Luther Burbank created the russet-colored Burbank potato. Russet Burbank potatoes formed the basis for a new potato-growing industry that developed in the northeastern states in the late nineteenth century (Bohl and Johnson 2010). Maine became the largest producer of potatoes in the 1920s and held that distinction until Idaho passed Maine to become the largest producer in the 1960s. Nearly all of Maine's potatoes are grown in Aroostook County in the far northeastern corner of the state (Maine Potato Board 2013).

Idaho's rise as the leading producer of potatoes coincides with a major shift in the crop's utilization. The invention of frozen French fries, a process asso-ciated with J. R. Simplot who was one of Idaho's largest potato growers, was instrumental in building the Idaho potato industry. In 1967 Simplot began supplying McDonald's fast-food restaurants with the new frozen potato prod-uct which was manufactured from Idaho-grown potatoes. By 1970 processed potatoes had passed table stock as the major use made of potatoes. At present table stock potatoes make up 26% of the total, chips and dehydrated products have a similar share, and frozen French fries account for 43% of potato con-sumption (National Potato Council 2015). Idaho grows about 30% of the U.S. total and the Columbia Basin of Washington and Oregon produces another 16%. Wisconsin, Minnesota, and North Dakota produce quantities of red potatoes which supply both fresh market and chipping uses.

Typical of western irrigated districts, the Columbia Basin of Washington and Oregon is a highly diversified producer of fruit and vegetable crops (Table 8.3). In addition to its large potato crop, the region grows roughly one-third of the onions produced in the United States, 85% of the pears, 30% of the green

Table 8.2   Concentration of vegetable crops, 2012

Crop	Half the crop acreage is in one or two counties in:
Artichokes	California
Broccoli	California
Brussels sprouts	California
Honeydew melons	California
Carrots	California
Cauliflower	California
Celery	California
Garlic	California
Lettuce	California, Arizona
Radishes	Florida
Spinach	California, Arizona

Crop	Half the crop acreage is in three or four counties in:
Asparagus	Michigan, California
Beets	New York, California
Cantaloupe	California, Arizona
Herbs	Texas, California, New Jersey
Parsley	California, Texas
Tomatoes	California

Crop	Half the crop acreage is in five to fifteen counties in:
Cabbage	Texas, Georgia, California, Wisconsin, New York
Cucumbers	Michigan, Wisconsin, California, Texas, Florida, Georgia
Eggplant	California, New Jersey, Hawaii, Florida, Georgia, Connecticut
Onions	California, Washington, Oregon, Texas, Idaho, Georgia
Peas	Oregon, Minnesota, Washington, Wisconsin
Peppers, Bell	Florida, California
Peppers, Chile	California, New Mexico
Potatoes	Idaho, Washington, Wisconsin, North Dakota, Maine, California
Snap Beans	Wisconsin, Florida, Oregon, New York, Texas, Illinois

Crop	Half the crop acreage is in more than fifteen counties in:
Watermelon	Texas, Indiana, California, Georgia, Florida, Arizona, Missouri
Pumpkin	Illinois, California, Texas, Pennsylvania, Virginia, New York, Michigan, Colorado, Delaware, New Jersey, Oregon, Ohio
Squash	Florida, Michigan, New York, California, New York, New Jersey, Oregon, Georgia, Arizona, Texas, Washington, Massachusetts
Sweet Corn	Washington, Minnesota, Wisconsin, Florida, Georgia, California, Idaho, Delaware

Source: USDA Census of Agriculture

peas, and 15% of the sweet corn. All of this is in addition to its 48% share of U.S. apple production. Construction of Grand Coulee Dam on the Columbia River during the 1930s made this massive irrigation district possible.

Tomatoes are the second most widely grown vegetable crop in the United States, after potatoes. Although more than 1,800 U.S. counties reported at

Table 8.3   Major vegetable regions and crops, 2012.

Region	Acres	Major Crops
Columbia Basin	339,299	Potatoes, Sweet Corn, Onions, Sweet peas
Southern San Joaquin Valley	310,843	Tomatoes, Carrots, Potatoes, Onions, Cantaloupe
Snake River Plain	254,472	Potatoes
Northern San Joaquin and Sacramento Valleys	195,913	Tomatoes, Sweet Corn, Cantaloupe
Imperial/Coachella Valleys and Colorado River	191,071	Lettuce, Cauliflower, Carrots
South Florida	125,369	Tomatoes, Sweet Corn, Snap Beans
New York/Lake Ontario	80,659	Snap Beans, Sweet Corn, Potatoes, Cabbage
Georgia-Florida Coastal Plain	74,250	Potatoes, Sweet Corn, Watermelon, Greens
Southern New Jersey/Delmarva	50,347	Sweet Corn, Potatoes, Tomatoes, Cucumbers
Lower Rio Grande Valley	38,323	Watermelon, Onions, Cabbage

Source: USDA Census of Agriculture

least some commercial tomato production in 2012, California produces 75% of the crop and Florida produces 10%. Florida is a major source of winter season fresh-market tomatoes and has about the same fresh-market acreage as California. But California has a 96% share of processed tomato production, which is the most important use made of the crop.

California, once the home of giant ketchup factories, has seen changes in the tomato processing industry. Major-brand food companies still receive and process a portion of the California tomato crop, but intermediate processors have now taken over much of that business. The intermediate companies harvest fresh tomatoes; convert them to paste, purees, ketchup, and diced and crushed products; and then pack them in 55-gallon drums and 300-gallon bins for delivery to the brand-label companies which might be located in California or anywhere else in the world (The Morning Star Co. 2015).

SUMMER VEGETABLES IN THE UPPER MIDWEST

Various influences of the physical environment have favored the production of fruits and vegetables in the Midwest during the warm summer season. One of the Great Lakes' moderating influences on climate is to decrease the likelihood of very cold winter temperatures, which has a beneficial effect on fruit trees. Western Michigan is the nation's largest source of tart cherries and the counties along Lake Michigan also produce crops of sweet cherries, tame blueberries, and apples.

An even greater variety of summer vegetables comes from Michigan, Wisconsin, and Minnesota (Figure 8.4). Long summer days at this latitude, combined with warm to hot daytime temperatures and cool nights, are excellent for crops such as sweet corn, beans, cucumbers, peas, and potatoes. The Upper Midwest has soils derived from glacial materials which have high nutrient levels and large organic matter content. Two of the best soils for growing vegetables are the black-colored, organic-rich "muck lands" and the freely drained sandy soils of former glacial lake bottoms, both of which are widespread from Michigan to North Dakota.

Nearly all summer vegetables are grown on contract. Farmers who grow vegetables might also be dairy farmers or cash-grain farmers or even part-time farmers who have cropland acres available. They sign contracts with a food processing company which takes charge of planting and harvesting the acreage, sometimes for more than one crop. Canning factories and frozen-food plants, scattered around Wisconsin and Minnesota, receive the harvests as they are trucked in from the fields. The green pea crop is generally harvested early in the summer, followed by green beans, then sweet corn. The canning factories are idle buildings most of the year but they come alive for each seasonal harvest. Maturation dates of the crops move from south to

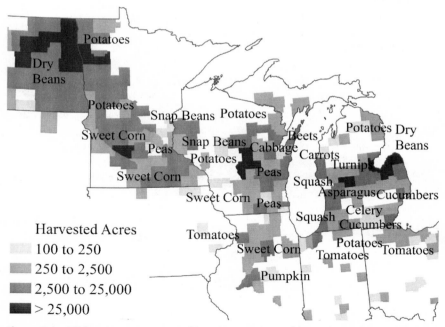

Figure 8.4 Midwest summer vegetables. *Source*: Created by authors, data from USDA Census of Agriculture 2012.

north, as do the harvests and the operations of the canning factories (Frederic 2002).

Harvested cucumbers go to pickle factories, harvested cabbage is carried to sauerkraut factories, while other crops are marketed fresh. Root vegetables such as turnips, parsnips, rutabagas, kohlrabi, and carrots are traditional north European crops grown where summers are mild, growing seasons are short, and daylight hours are long. The same mixture of crops is produced in Michigan, which approximates those same growing conditions.

Sweet corn is marketed fresh in the summer, although much of the sweet corn grown in the Upper Midwest moves to canning factories or frozen food plants. On a national basis, sweet corn is generally grown along with other vegetable crops rather than in the same areas that grow corn for grain, but occasionally the two crops coincide. The farmers of Renville County, Minnesota, harvest more than 20,000 acres of sweet corn in a typical summer which makes their county rank among the top three or four sweet-corn-producing counties in the United States. But in Renville County sweet corn occupies less than 10% as many acres as the cash-grain corn crop requires. Sweet corn is a seasonal specialty crop, grown in Florida in the winter and in the northern states during the summer months. Despite its physical resemblance to grain corn, the two have little to do with each other.

Sandy soils are favored for the pumpkin crop of the Illinois River valley and the watermelon crops of the lower Wabash Valley. Illinois's pumpkins are consumed by nearby canning factories that produce pumpkin-pie filling while the Wabash Valley watermelons are marketed fresh. Sandy soils also support a large potato acreage in central and northern Wisconsin. All of these crops are widely adapted to Midwestern environments, but the presence of a specific environmental advantage such as freely draining soils can attract the crop to a particular site.

North Dakota is the largest producer of dry edible beans, including kidney, pinto, white, and great northern varieties. Dry beans and potatoes can be fit into the short summer growing season of the Upper Midwest and typically they are grown north of all other vegetable specialties. The main use of dry beans is in canned form. With the addition of the green and red chilies grown in the Rio Grande Valley of New Mexico, they are essential ingredients for Mexican-food processing businesses in the United States.

INCREASED CONSUMPTION OF FRESH
FRUITS AND VEGETABLES

Per capita consumption of all fruits and vegetables has grown slowly and erratically in the United States over the past four decades. The fresh component of the total has increased from 40% to 45% in the early 1970s to around 45%–50% today. Imports account for about half the growth in consumption of fresh fruits and a quarter of increased fresh vegetable consumption. The North American Free Trade Agreement has greatly stimulated fresh produce imports to the United States from Mexico, with imports increasing from around $1 billion a year in the early 1990s to about $4 billion today. The desirability of increasing the fresh fruit and vegetable content of the American diet is commonly discussed, but there is no marked trend in that direction (Dong and Lin 2009). The major contribution made by the fresh fruit and vegetable imports is not in the total amount as much as in the enhanced seasonal availability of the products. Blueberries, strawberries, raspberries, and other fruits and vegetables that once had a limited seasonal availability are now available as fresh market offerings for much of the year because of the imports.

REFERENCES

Bohl, W. H. and S. B. Johnson (eds.). 2010. Commercial Potato Production in North America. Orono ME: Potato Association of America. [http://potatoassociation.org/wp-content/uploads/2014/04/A_ProductionHandbook_Final_000.pdf].

California Department of Food and Agriculture. 2015. *California Grape Acreage Report, 2014.* Sacramento. [http://www.nass.usda.gov/Statistics_by_State/California/Publications/Grape_Acreage/201504gabtb00.pdf].

Dong, D., and B.-H. Lin. 2009. *Fruit and Vegetable Consumption by Low-Income Americans. Would A Price Reduction Make A Difference?* Economic Research Report No. 70. USDA. Economic Research Service. [http://www.ers.usda.gov/media/185375/err70.pdf].

Ferrier, P. 2014. Imports of Many Fruits and Vegetables Dominated by Few Source Countries. *Amber Waves.* USDA Economic Research Service [http://www.ers.usda.gov/amber-waves/2014-august/imports-of-many-fruits-and-vegetables-dominated-by-few-source-countries.aspx#.VbpHcvlVhBc].

Florida Citrus Mutual. 2015. *Citrus Industry History.* [http://www.flcitrusmutual.com/citrus-01/citrushistory.aspx].

Frederic, P. B. 2002. *Canning Gold; Northern New England's Sweet Corn Industry: A Historical Geography.* Lanham, MD: University of Press of America.

Geisler, M. 2012a. *Blackberries.* USDA. Agricultural Marketing Resource Center. [http://www.agmrc.org/commodities__products/fruits/blackberries/].

_____. 2012b. *Raspberries.* USDA. Agricultural Marketing Resource Center. [http://www.agmrc.org/commodities__products/fruits/raspberries/].

Huang, S. and K. Huang. 2007. *Increased U.S. Imports of Fresh Fruits and Vegetables.* FTS-328-01. USDA. Economic Research Service [http://www.ers.usda.gov/media/187841/fts32801_1_.pdf].

Maine Potato Board. 2013. *A Review of the Industry.* Presque Isle: Maine Potato Board. [http://www.mainepotatoes.com/page/956-729/maine-potato-industry-reports].

Michigan Apple Committee. 2015. *The Michigan Apple Crop 2014.* http://www. michiganapples.com/News/Michigan-Apple-Facts.

The Morning Star Company. 2015. Website: http://www.morningstarco.com/.

National Potato Council. 2015. *Potato Facts.* [http://www.nationalpotatocouncil.org/ potato-facts/].

Perez, A., and K. Plattner. 2015. *Fruit and Tree Nuts Outlook.* FTS-358. USDA Economic Research Service. [http://usda.mannlib.cornell.edu/usda/current/FTS/ FTS-03-27-2015.pdf].

Plattner, K. 2014. *Fruit and Tree Nuts Outlook: Economic Insight. Fresh-Market Limes.* USDA. Economic Research Service. FTS-357SA. http://www.ers.usda.gov/ media/1679187/fresh-market-limes-special-article.pdf.

USDA. Economic Research Service. 2015. *Fruit and Tree Nuts.* [http://www.ers. usda.gov/topics/crops/fruit-tree-nuts/trade.aspx#Fruit].

USDA. National Agricultural Statistics Service. 2015. *Noncitrus Fruits and Nuts, Preliminary Report 2014.* [http://www.nass.usda.gov/Publications/Todays_Reports/ reports/ncit0115.pdf].

Zahniser, S., S. Angadjivand, T. Hertz, L. Kuberka, and A. Santos. 2015. *NAFTA at 20: North America's Free-Trade Area and Its Impact on Agriculture.* USDA. Economic Research Service. WRS-15-01. [http://www.ers.usda.gov/media/1764579/ wrs-15-01.pdf].

Chapter 9

Organic Farms and Organic Food

The organic sector differs from the rest of American agriculture not so much in the items it offers for sale but rather in terms of how and according to what standards those items are produced. The growth of organic farming and the demand for organic products has been one of the most significant developments in food production over the past 25 years. To quote the USDA, "Organic food is produced by farmers who emphasize the use of renewable resources and the conservation of soil and water to enhance environmental quality for future generations." Organic food "is produced without using most conventional pesticides." It avoids "fertilizers made with synthetic ingredients," and uses no genetically modified organisms (GMOs). Organic meat, poultry, eggs, and dairy products come from animals that are given no antibiotics or growth hormones. Organic production is characterized by the USDA as integrating "the parts of the farming system into an ecological whole," by using materials and practices that "enhance the ecological balance of natural systems" (USDA Agricultural Marketing Service 2015a).

Growth of the organic sector has come largely from a conversion of conventional farms to organic farms as they modify their operations to meet the USDA's Certified Organic requirements. In 2002, the first year that the Certified Organic label was used, 11,998 farms qualified. That number advanced slowly to a total of 14,326 farms in 2012, although it had been as high as 18,211 in the 2007 Census of Agriculture. The expansion of organic farming has taken place at a time when the total number of American farms is decreasing but the value of organic production per farm is increasing (Greene 2013). Still, organic farms account for a small share (0.67%) of all farms in the United States. Even though organic food is widely available in supermarkets, farmers markets, and food coops, it does not come close to rivaling the public's consumption of conventional foods.

FEDERAL ORGANIC STANDARDS

Prior to 1990 there was little consensus about what constituted "organic." Pressure from farm organizations, food retailers and wholesalers, environmentalists, and consumer groups led to congressional passage of the Organic Foods Production Act (OFPA) of 1990, or Title 219 of the 1990 farm bill. Authored by U.S. senator Patrick Leahy—true to his organic-conscious constituents in the state of Vermont—the OFPA established a basis for comprehensive federal regulation of organic farms and food. It made the USDA responsible for establishing national standards for organic production and marketing, for assuring consumers that organically produced products meet those standards, and for facilitating commerce in organically produced products (U.S. Secretary of Agriculture 2013).

Beginning in 2002 producers who marketed commodities labeled as organic were required to be certified by a state agency or private organization as being in compliance with USDA organic standards. Those who met the requirements were allowed to display the "Certified Organic" label on their products and in their advertising. Farms with less than $5,000 in gross sales from organic products were exempt from the requirement, leading to the two categories, Certified Organic and Certified Exempt. The cost to the farmer of the certification procedure ranges from $750 to thousands of dollars, depending on the size of the farm. Since 2008 part of the cost is eligible for reimbursement by the federal government. Organic certification is available to any farm, food handler, or sales organization in the world that wishes to be certified as organic by the USDA. The network of certifying agents is worldwide and operates to include organic shipments between countries as well as within the United States (Baier 2012).

At present about five dozen accredited organic certification programs are in operation within the United States; twenty-two of them are in the hands of state or county governments and about forty are private firms. The accrediting firms themselves are subject to inspection to guarantee they are operating as intended. Visits to the organic farm, review of the farm's organic plan for operation, and interviews with neighbors are all included as steps in the review process before a farm can be certified. Organic certification is a designation that applies to specified farm products, not necessarily to entire farms. A farmer producing organic wheat, for example, might well produce conventional wheat as well, or might engage in a variety of other farm businesses. According to the 2012 Census of Agriculture, more than half of all farms selling one or more Certified Organic products also produced conventional crops or livestock.

The 1990 OFPA legislation also established a fifteen-member National Organic Standards Board which includes environmentalist, farmer/grower, consumer, and retailer members who advise the National Organic Program on substances and practices that should be either allowed or prohibited in organic farming and food distribution. The Board, which is appointed by the Secretary of Agriculture, only makes recommendations and does not set policy (USDA, Agricultural Marketing Service 2015d).

The standards also apply to organic food products sold to the public. Products labeled "organic" must consist of at least 95% organically produced ingredients; products labeled "made with organic ingredients" must contain at least 70% organic ingredients; and products with less than 70% organic ingredients cannot use the term "organic" anywhere on the principal display panel (USDA Agricultural Marketing Service 2015b).

ORGANIC FARMS

Recent federal policy has sought to increase the size of the organic sector by increasing the number of Certified Organic farms by 25% (Greene 2014). The failure of organic farms to show more than the modest increase in numbers may be attributed in part to the costs and administrative burden of the certification procedure itself. But despite the slow growth in the number of organic farms, total productivity of the sector has grown from $393 million in gross sales in 2002 to more than $3 billion in 2012. Just since 2007 the value of organic sales from farms has increased 84%.

At present, half of all organic farms have sales less than $35,000 per year, while two-thirds of all organic sales come from the largest 8% of farms, a concentration similar to that which has long characterized conventional agriculture. These observations suggest that the organic sector has performed in much the same way that conventional agriculture has over the past decades. Most of the sales are accounted for by the larger operations, whether they are organic or conventional.

The difference remains even if Certified Organic status is removed from the analysis. In 2014 the USDA conducted a survey of 163,675 farms that marketed food locally, either through direct-to-consumer sales or through intermediate channels (Low and others 2015). Eighty-five percent of these farms had a gross income less than $75,000 and accounted for only 13% of local food sales. In contrast, local food farms with a cash income greater than $350,000 accounted for only 5% of the farms but produced 67% of the sales (see also Chapter 1). The small farms, while both numerous and widespread, do not produce much of the total.

Organic and non-organic farms do not vary widely in terms of their acreages. The smallest farm operations, having fewer than ten acres, account for 17.5% of all organic farms but only 10.6% of non-organic. At the upper end of the scale, 3.9% of non-organic farms and an identical 3.9% of organic farms have more than 2,000 acres (Census of Agriculture 2012).

The most common types of organic farms are those that concentrate on the production of vegetables, fruit, or dairy products. Organic vegetable farms make up about 7% of the total number of vegetable farms in the United States, while organic dairy farms account for about 5% of all farms specializing in dairy products. These averages vary substantially from state to state. Seventeen percent of Vermont's dairy farms and nearly 25% of its vegetable farms are in the Certified Organic category. Maine and New Hampshire similarly have high percentages of vegetable and dairy farms in the organic category, while California, Wisconsin, Oregon, and Washington have moderate percentages. Elsewhere, the percentages are very low.

The national map of organic farms shows several high-density clusters separated by vast expanses where organic farms are comparatively rare (Figure 9.1). In the Northeast, organic farms have a strong presence in New England, New York, and eastern Pennsylvania. A second cluster appears in the Upper Mississippi Valley of Wisconsin, Minnesota, and Iowa. The largest region of organic farm concentration stretches the length of the Pacific Coast from southern California to Washington.

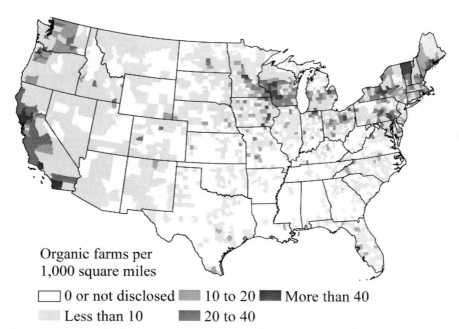

Organic farms per
1,000 square miles

☐ 0 or not disclosed ▨ 10 to 20 ■ More than 40
   Less than 10 ▧ 20 to 40

Figure 9.1   Organic farms, 2012. *Source*: Created by authors, data from USDA Census of Agriculture 2012.

Because many organic farms produce for local markets, it is not surprising that they are clustered in the more densely populated parts of the country and also are absent from areas where urban populations are small. Wisconsin, New York, and New England are dairy producers and many dairy farms there are Certified Organic. Washington, Oregon, and California are leading producers of fruits and vegetables, and this association would tend to increase their numbers of organic farms as well.

Grain and oilseed farms, cattle grazing operations, feedlots, and poultry farms are much less common among the ranks of organic farms. The comparative lack of organic producers in the South and the Great Plains, where those types of farm specialties are most common, may be partly explained by that circumstance. Organic crops tend to be grown in the same places that those receiving conventional treatment are produced. As an example, in 2011 the United States produced 132,000 acres of organic soybeans (0.2% of the national soybean acreage), roughly 36% of which was grown in Iowa, Minnesota, and Michigan, all three states being major conventional soybean producers as well (USDA, Economic Research Service 2013).

Other factors associated with concentrations of organic farms include religious and cultural affiliations. Old Order Mennonite and Old Order Amish people tend to avoid technologies they believe are destructive to the future of their communities. While some Amish and Mennonite practices—such as limited use of electricity—would not be directly associated with organic production, other avoidances such as chemical fertilizers, pesticides, and herbicides would place them in compliance with Certified Organic rules. Organic farm concentrations in southeastern Pennsylvania, north-central Ohio, northeastern Indiana, western Wisconsin, and eastern Iowa correspond with areas of Old Order Amish or Old Order Mennonite settlement.

ORGANIC FOOD PRODUCTION

Viewing the organic sector from the perspective of production (rather than just the number of farms) reveals a somewhat different picture. Measured by total farm sales organic agriculture is heavily concentrated in only a limited number of U.S. counties (Figure 9.2). Sixty percent of the value of organic production comes from the West Coast states, with California alone contributing more than 44% of the national total (Greene and Ebel 2012). The Upper Mississippi Valley produces another 8%, the same as New England and New York. Less than one fourth of U.S. organic production comes from outside these three clusters.

The strong concentration of organic production in those three areas is partly a function of their crop and livestock specialties. Organic apples are

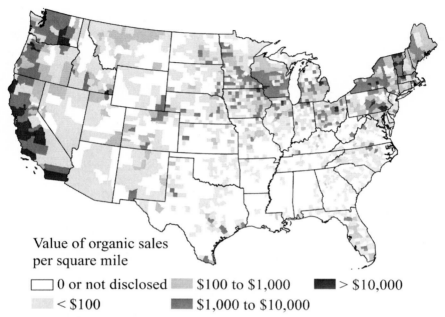

Figure 9.2 Organic farm sales, 2012. *Source:* Created by authors, data from USDA Census of Agriculture 2012.

an example. The state of Washington has long led all other states in the production of apples, and for organic apples Washington has an even larger, 90% share of fresh-market production. Hot, dry conditions in Washington's irrigated apple valleys keep away many insects and diseases that threaten tree crops. This means the organic producer's avoidance of chemical pesticides and insecticides is less of a limitation in Washington than it would be in more humid apple-producing states such as Michigan.

Organic citrus comes from California (62%) and Florida (35%), the same states that produce most conventionally grown citrus fruit. California's share is much larger than Florida's for organic production, again reflecting the advantage of a dry summer climate that reduces the need for controlling the insect species which organic farmers have a smaller variety of options to combat. California's list of firsts goes on: the state produces 60% of all organic green and mixed vegetables, 60% of total organic fruits and nuts, and 87% of organic grapes. California's pre-eminent position is not just confined to crops. The state produces $6.9 billion of organic milk each year, more than 40% greater than Wisconsin; and its farms and ranches produce $1.34 billion of organic beef every year, the largest quantity of any state. In 2011, nearly 80,000 head of organic cattle were sold for slaughter in the United States, about one-fourth of which were processed in California with the rest widely distributed over the organic-producing counties of the nation. California is the nation's largest agricultural producer in total and even more overwhelmingly the largest organic producer.

ORGANIC FOOD, LOCAL FOOD, AND COMMUNITY-SUPPORTED AGRICULTURE

The preceding discussion makes it clear that much of the organic food consumed in the United States must travel a substantial distance before it is consumed, unless the point of consumption happens to be on the West Coast or in a few scattered enclaves elsewhere. Certified Organic food and locally grown food obviously are not the same thing. The impulse to grow at least some of one's own food, or to purchase food from a local farmer, is far more widespread in the population than is the production of food that meets federal organic standards (Toler and others 2009).

Some certifying programs have been created as an alternative to the USDA Certified Organic label. Certified Naturally Grown (CNG), based in Brooklyn, New York, has more than 700 farm members at present, which is only about 5% as many as are Certified Organic. CNG is characterized as a nonprofit, alternative eco-labeling program for small farms that follow USDA organic methods but are not part of the USDA program. CNG certification is less involved than the USDA procedure and inspections of CNG farms are carried out only by other CNG farmers (Certified Naturally Grown 2015). A recent study has shown that consumers are willing to pay a substantial premium for the "certified organic" label, but that a similar benefit is absent from farms that advertised being CNG (Conolly and Klaiber 2014). The state of Georgia currently has the largest number of CNG farms (122), while Pennsylvania, New York, and Virginia have several dozen each.

The term "organic" represents an official U.S. government–authorized designation (Kremen and others 2013). Its unlawful application to other products does carry penalties, but American consumers use "organic" to denote a variety of food-producing operations that are typically small in size and promise not to use agricultural chemicals (Dimitri 2011). Research has shown that consumers are satisfied with a food source being "local" if it originates in the same state, which at least provides a rough definition of the term (Darby and others 2008). The 163,675 farms that the USDA counted as marketing food locally in 2012 represented 7.8% of all farms in the United States. Some of those many enterprises probably are perceived as being "organic" even though they are not Certified Organic, although farm owners are careful not to enter such a claim if their operations do not meet the USDA standards.

Another designation that has grown in a fashion somewhat parallel to organic certification is the Community Supported Agriculture movement (CSA). Individual consumers purchase shares in a CSA early in the growing season in exchange for produce that comes from the farm in later months. The risks of bad weather or crop failure are shared by the farmer and the consumer in this arrangement. The customer typically receives a box of produce on a regular basis as the season progresses from one crop to the next. CSAs advertise their practices, including their policies on the use of agricultural chemicals, so that share purchasers can be informed about what they are getting.

Choices of produce are limited, dictated by the season in which the crop matures, the same as is true for a home garden. Shares typically cost from $50 to $750, depending on the length of the season. Some CSAs allow shareholders to work on the farm in exchange for a reduced payment, while others forbid it. Shareholders who do not want the offering that the CSA provides at a given time have the option of declining it, although typically they receive no refund for what has been forgone.

CSAs began to appear in New England during the 1980s after the concept was introduced from Europe (McFadden 2004). Informal histories of the CSA maintain that it originated in the 1920s in a series of lectures given by an Austrian philosopher, Rudolf Steiner (1861–1925). Steiner described an approach called "biodynamic agriculture" which was intended to instruct farmers how "to influence organic life on earth through cosmic and terrestrial forces" (Chalker-Scott n.d.). Although the spiritual dimensions of Steiner's ideas have been criticized, his concept of the producer-consumer association, a local-level collective in which producers and consumers are linked by their mutual interests, is readily seen as the organizing principle behind the CSA. The first two CSA farms, which were started in New England in 1986, are still in operation. In 2015 the USDA Agricultural Marketing Service directory of CSAs listed 661 operations, with the largest numbers in New York (50), Wisconsin (41), and Michigan (34) (USDA, Agricultural Marketing Service 2015c).

THE PRICES AND COSTS OF ORGANIC FOOD

The higher prices commanded by organic products in the marketplace sometimes are explained in terms of supply and demand: organic food products are in short supply, hence their price is bid up. It seems likely that such a condition would not exist for long before more producers would enter the market, however, which would provide an additional supply that would bring the price down. Another explanation is that organic products cost more to produce and hence they carry a higher price which the consumer must pay.

If organic products were not valued by consumers, they would be unable to command such a position above conventional products.

As to why organic products are valued above the conventional, three explanations are generally offered (Greene and others 2009). One is the perceived health and safety benefits of organic food (Duram 2005, Hansen 2010). Crops raised with the use of herbicides, insecticides, and fungicides are considered dangerous to human health by many people. Avoiding those products would then contribute to one's health and well-being. A growing segment of the national population favors banning anything considered not "natural" from food, which sometimes means that a less commonly available and hence more expensive ingredient must be substituted.

A second argument involves the role of fairness (Toler and others 2009). Small, family farms have difficulty competing with large farms because they are small and hence are unable to take advantage of the economies of scale that larger farms enjoy. Those who argue that this situation is unfair to the small farmer often believe it is conscionable to pay more money for a product grown or raised on a small, local farm. Although such farms are not necessarily also following organic certification rules, they generally employ a lower level of technology, often substitute family labor for capital, and purchase fewer inputs from outside the farm, much in the same spirit as an organic farm would. This view coincides with the more general concept of "fair-trade" marketing where higher prices are charged for a product with the promise that more money will be returned to the point of production.

A third rationale for higher prices on organic products is the role of "ecosystem services" that organic farms perform. Organic food costs more to purchase because ecosystem services represent a cost to the producer which is added to the price of food. Conventional agriculture, which is viewed as being much less concerned with its place in the ecosystem, offers products for sale at a cheaper price because the non-organic producer adds fewer ecosystem services (Greene and others 2009).

While practices such as minimum tillage, use of conservation buffers, sowing of cover crops, contour plowing, rotational grazing, and humane treatment of animals are also widely found on conventional farms, many are required by federal law under terms of the Certified Organic program. Farm management strategies, which are good for the environment, do not necessarily cost more and in many cases they also can be viewed as cost-cutting alternatives. National policy favors and supports farms that employ conservation strategies regardless of their organic certification status (Duram and Oberholtzer 2010).

The higher selling price of organic food in the marketplace begins at the farm and it persists through the wholesale and retail chains for a variety of products and commodities. A 2005 survey of dairy farms, for example, showed that organic dairy farms had operating and capital costs $5.65 to

$6.37 cwt higher than the conventional dairy farm, but the average price pre-mium for organic milk was $6.69 higher, which covered the additional costs (McBride and Greene 2009).

One of the requirements for organic certification of a dairy farm is pasture access for cows. Pasture use also substitutes for harvested forage and other, more expensive feed inputs, but it requires more labor and is limited by the amount of land available if pasture is used to provide most of the feed. Most organic dairies are smaller on the average than their conventional counter-parts and they rely on the pastures for grazing during the months when that is feasible. But the reliance on pasture may produce a net financial loss when the cost of unpaid family labor is factored in.

Organic dairies produce about 30% less milk per cow than conventional dairies and their operating and capital costs per cwt of production are higher on the average because the organic dairies are smaller. As organic dairy farms grow in size they take advantage of lower production costs with increased output and their practices tend to move closer to those of the conventional dairy farm. Organic dairy farming is attractive and it is growing. Organic milk accounted for 4.38% of total milk sales in the United States in 2012, more than double the percentage that was recorded in 2006.

Apples are one of the largest organically produced crops. In a 2007 sur-vey, the USDA found that 45% of organic apple growers are in the business primarily to increase farm income (Slattery and others 2011). The farm price of organic apples ranges from 60% higher to more than double the price of conventional apples. In 2012 the organic premium on Gala apples in the wholesale market was about 54% over the conventionally grown. Organic apples now account for 4.9% of total U.S. apple production and the percent-age is slowly growing.

Fresh vegetables have even higher organic price premiums. Looking at just the 2012 wholesale market prices at San Francisco and Atlanta (as reported by the USDA Agricultural Marketing Service), price premiums on organic russet potatoes, spinach, onions, Romaine lettuce, cauliflower, and carrots average above 200%—that is, more than twice the price of the conventional product (USDA Agricultural Marketing Service 2015e). The wholesale price premium on fruits, including strawberries, bananas, navel oranges, and Bartlett pears, ranges from 149% to 235%. Some of these fruits arrive fresh in the United States in various seasons of the year from foreign growers who have achieved USDA Certified Organic status. Certified Organic avocadoes produced in Mexico are one example of foreign suppliers in the organic marketplace.

There are equally strong price incentives for grain and oilseed farmers to shift at least part of their production into the organic mode. Although only a small fraction of U.S. soybean acreage is in organic production, organic soybeans bring a high price because of two other characteristics with which they are identified. Food-grade crops command higher prices because they are suitable for direct human consumption. Several products, including miso, tofu, natto, and soy milk, require food-grade soybeans for their preparation. Most food-grade soybeans are also non-GMO (not genetically modified), which means that they must be grown, processed, stored, and shipped apart from conventional soybeans.

The non-GMO issue has given rise to the IP (Identity Preserved) label, which means that the product is guaranteed not to have come in contact with GMOs. For soybeans and other grain crops, this generally means the commodity was produced under Certified Organic conditions and then shipped via sealed container to its final destination, no matter where that might be. In 2011, 7% of U.S. soybeans were transported via container, a portion of which consisted of shipments of food-grade soybeans to China, Taiwan, Indonesia, and Japan (McBride and others 2012). IP export soybeans can be loaded directly into the container in the field. The container is trucked to a rail or ocean terminal and then is forwarded to its destination aboard ship.

Organic soybean yields per acre are lower and costs of production are higher than for conventional soybeans. A 2006 USDA study showed that total economic costs were $6.20 higher for organic soybeans but in that year the price premium was $9.16 per bushel for organic (McBride and Greene 2013). Any price premium for food-grade soybeans would be added to the organic premium. Because USDA organic certification requires that cropland be free of non-organic influences for several years before an organic crop may be sold as such, producers of organic soybeans have less ability to take advantage of year-to-year fluctuations in the prices paid for their crops.

Organic wheat made up 0.68% of the total 2008 wheat acreage in the United States, a small fraction and comparable with soybeans. But unlike the soybean crop, which is roughly 95% GMO in the United States today, no GMO wheat is currently in production. Since there is no advantage to organic wheat as the non-GMO alternative to conventional wheat, growers who raise organic wheat rely mainly on the price incentive. Organic wheat yields are lower in part because of the difficulty in controlling weeds, but since no herbicide costs are involved in raising organic wheat its production costs are thereby lowered. In 2009 operating costs plus capital costs were $2 to $4 per bushel higher for organic wheat, but the price premium was $3.79, which was enough to make the organic alternative profitable for some farmers depending on their costs (McBride and others 2012). Montana, Utah, and Colorado account for about 40% of U.S. organic wheat acreage. IP status is also important in wheat and containerized shipping is often used, which adds substantially to transportation costs.

TRENDS IN ORGANIC PRODUCTION

Organic food production is growing in the United States, although at a slower rate than what current national policy favors. A recent USDA report noted, "Despite the potential for organic agriculture to improve the environmental performance of U.S. agriculture, the national standard is having only a modest impact on environmental externalities caused by conventional production methods because the organic adoption rate is so low." In other words, more organic farms would produce more ecosystem services (Greene 2014).

Urban agriculture—the idea that significant amounts of food can be produced in cities—is discussed in popular media but remains largely untested. Crop yields in urban and urban-fringe agricultural settings have been highly variable, indicating that food production in nontraditional contexts is likely to be risky (Wagstaff and Wortman 2015). The risk is even greater given that proponents of urban farming stress its potential for making better food available to low-income, inner city residents.

As with all commercial ventures, organic farming is ultimately driven and restrained by market forces (Dimitri and Oberholtzer 2009). Given the bargain prices of most conventionally produced food items offered for sale in the United States, it would appear that higher-priced organic alternatives will continue to appeal to a segment of the total market but will face consumer resistance because of the price difference.

REFERENCES

Baier, A. H. 2012. *Organic Certification of Farms and Businesses Producing Agricultural Products*. USDA Agricultural Marketing Service. [http://www.ams.usda.gov/AMSv1.0/getfile?dDocName=STELPRDC5101547].

Census of Agriculture. 2012. *Characteristics of All Farms and Farms with Organic Sales. Certified Naturally Grown*. 2015. [https://www.naturallygrown.org/].

Chalker-Scott, L. n.d. *The Myth of Biodynamic Agriculture*. [http://puyallup.wsu.edu/wp-content/uploads/sites/403/2015/03/biodynamic-agriculture.pdf].

Conolly, C., and H. A. Klaiber. 2014. Does Organic Command a Premium When the Food is Already Local? *American Journal of Agricultural Economics* 96(4): 1102–1116.

Darby, K., M. T. Batte, S. Ernst, and B. Roe. 2008. Decomposing Local: A Conjoint Analysis of Locally Produced Foods. *American Journal of Agricultural Economics* 90(2): 476–486.

Dimitri, C. 2011. Use of Local Markets by Organic Producers. *American Journal of Agricultural Economics* 94(2): 301–306.

Dimitri, C., and L. Oberholtzer. 2009. *Marketing U.S. Organic Foods: Recent Trends From Farms to Consumers*. USDA Economic Research Service. Economic Information Bulletin EIB-58.

Duram, L. 2005. *Good Growing: Why Organic Farming Works.* Lincoln: University of Nebraska Press.

Duram, L. A., and L. Oberholtzer. 2010. A Geographic Approach to Examine Place and Natural Resource Use in Local Food Systems. *Renewable Agriculture and Food Systems* 30(1): 99–108.

Greene, C. 2013. Growth Patterns in the U.S. Organic Industry. *Amber Waves.* USDA Economic Research Service. [http://www.ers.usda.gov/amber-waves/2013-october/growth-patterns-in-the-us-organic-industry.aspx#.Vbj-SflVhBc].

_____. 2014. Support for the Organic Sector Expands in the 2014 Farm Act. *Amber Waves.* USDA, Economic Research Service. [http://www.ers.usda.gov/amber-waves/2014-july/support-for-the-organic-sector-expands-in-the-2014-farm-act.aspx#].

Greene, C., C. Dimitri, B-H. Lin, W. McBride, L. Oberholtzer, and T. Smith. 2009. *Emerging Issues in the U.S. Organic Industry.* USDA, Economic Research Service. Economic Information Bulletin Number 55.

Greene, C., and R. Ebel. 2012. Organic Farming Systems. In *Agricultural Resources and Environmental Indicators, 2012 edition,* edited by C. Osteen, J. Gottlieb, and U. Vasavada. 2012. USDA, Economic Research Service, Economic Information Bulletin Number 98, pp. 37–40.

Hansen, A. L. 2010. *The Organic Farming Manual.* North Adams MA: Storey Publishing.

Kremen, A., C. Greene, and J. Hanson. 2013. *Organic Produce, Price Premiums, and Eco-Labeling U.S. Farmers Markets.* USDA Economic Research Service. VGS 301–01 [http://www.ers.usda.gov/media/269468/vgs30101_1_.pdf].

Low, S. A., A. Adalja, E. Beaulieu, N. Key, S. Martinez, A. Melton, A. Perez, K. Ralston, H. Steward, S. Suttles, S. Vogel, and B. B. R. Jablonski. 2015. *Trends in U.S. Local and Regional Food Systems: A Report to Congress.* USDA, Economic Research Service. Administrative Report AP068 [http://www.ers.usda.gov/publications/ap-administrative-publication/ap-068.aspx].

McBride, W. D. and C. Greene. 2009. *Characteristics, Costs, and Issues for Organic Dairy Farming.* USDA. Economic Research Service, ERR-82. [http://www.ers.usda.gov/publications/err-economic-research-report/err82.aspx].

_____. 2013. *Organic Data and Research from the ARMS Survey: Findings on Competitiveness of the Organic Soybean Sector.* USDA, Economic Research Service [http://handle.nal.usda.gov/10113/58108].

McBride, W. D., C. Greene, M. B. Ali, and L. F. Foreman. 2012. *The Structure and Profitability of Organic Field Crop Production: The Case of Wheat.* Paper presented at Agricultural and Applied Economics Association meeting, Seattle WA. [http://ageconsearch.umn.edu/bitstream/123835/2/AAEA%20paper-organic%20wheat.pdf].

McFadden, S. 2004. *The History of Community Supported Agriculture, Part I.* [http://newfarm.rodaleinstitute.org/features/0104/csa-history/part1.shtml].

Slattery, E., M. Livingston, C. Greene, and K. Klonsky. 2011. *Characteristics of Conventional and Organic Apple Production in the United States.* FTS-347-01. USDA, Economic Research Service. [http://www.ers.usda.gov/media/118496/fts34701.pdf].

Toler, S., B. C. Briggeman, J. L. Lusk, and D. C. Adams. 2009. Fairness, Farmers Markets, and Local Production. *American Journal of Agricultural Economics* 91(5): 1272–1278.

USDA, Economic Research Service. 2013. *Certified Organic Grain Crop Acreage by State. 2011.* [http://www.ers.usda.gov/data-products/organic-production.aspx#25766].

USDA, Agricultural Marketing Service. 2015a. National Organic Program. [http://www.usda.gov/wps/portal/usda/usdahome?contentidonly=true&contentid=organic-agriculture.html].

_____. 2015b. Certified Organic. [http://www.ams.usda.gov/AMSv1.0/nop].

_____. 2015c. Community Supported Agriculture Directory Search. [http://search.ams.usda.gov/csa/].

_____. 2015d. National Organic Standards Board. [http://www.ams.usda.gov/AMSv1.0/getfile?dDocName=STELPRDC5101547].

_____. 2015e. Organic Prices. 2012. [http://www.ers.usda.gov/data-products/organic-prices.aspx].

U.S. Secretary of Agriculture. 2013. *USDA Departmental Guidance on Organic Agriculture, Marketing and Industry* [http://www.usda.gov/documents/usda-departmental-guidance-organic-agriculture.pdf].

Wagstaff, R. K., and S. E. Wortman. 2015. Crop Physiological Responses across the Chicago Metropolitan Region: Developing Recommendations for Urban and Peri-Urban Farmers in the North-Central U.S. *Renewable Agriculture and Food Systems* 30(1): 8–14.

Chapter 10

The Reserved Lands

Unlike urbanized areas, where most land is used intensively, agricultural lands are open and provide services in addition to their principal use as providers of food, fiber, and fuel. At the same time an acre of farmland is yielding 200 bushels of corn it may also serve as habitat for wildlife. Wild game such as white-tailed deer and ring-necked pheasants often rely on the cover of corn stalks when seeking shelter from the weather, especially during winter (Laingen 2011). The remains of crop plants left in the ground help reduce soil erosion and after they have decayed they increase the organic-matter content of the soil. These kinds of functions, often grouped under the heading "ecosystem services," are very much a part of the agricultural landscape (Reganold and others 2011).

Providing ecosystem services can itself be the principal use made of a tract of rural land. Over the past century designated lands have been idled through government-regulated cropland retirement programs, usually after they have been declared marginal for crop production. The cropland's marginality may be such that the farmer can decide whether to temporarily or permanently retire the land from production. Some government programs retire lands on an annual basis, while others require lands to remain out of production for a decade or longer.

Because many of these programs have been initiated to preserve soil and water quality, most programs retire croplands for at least a decade. At the end of a ten-year contract, the landowner may elect to keep the land retired or resume production. The agricultural market and whether more money can be made by farming the land or by keeping it in retirement typically guide the landowner's decision. While arguing the efficacy of conservation programs, critics have questioned if such taxpayer-subsidized initiatives can indeed exist side by side with production agriculture and effectively remove both overproduction and environmental degradation from the American agricultural landscape (Leathers and Harrington, 2000; Gersmehl and Brown, 2004).

RESERVED LANDS IN THE 1930s AND 1940s

Ever since 1933 the U.S. Congress has determined how taxpayer dollars should be used to support agricultural programs through various appropriation acts known as farm bills. The first farm bill was the Agricultural Adjustment Act (AAA) of 1933 which was part of President Franklin D. Roosevelt's first New Deal. One out of every four Americans lived on a farm at that time and gross farm incomes had dropped by over 50% as the Great Depression worsened (Cain and Lovejoy 2004). Through the use of price supports farmers voluntarily removed cropland from production in order to control the supply of food, which in turn drove up crop prices. In 1936 the U.S. Supreme Court ruled that the AAA was unconstitutional because consumers were being forced to pay a tax that helped only one segment of society: farmers. Shifting the focus from crop production to the land, Congress then passed the Soil Conservation Act of 1936 and created the Soil Conservation Service (SCS). The SCS was passed during the Dust Bowl years of the 1930s and it eventually distributed funds to farmers who implemented soil conservation programs.

In 1936 the Soil Conservation and Domestic Allotment Act provided funding through an Agricultural Conservation Program (ACP) that paid farmers to replace crops such as corn, cotton, and wheat that caused excessive soil loss with grasses and legumes that conserved the soil. Again the purpose was to reduce crop surpluses and raise prices. Farmers enrolled their marginal lands in conservation programs and used the government payments they received to spread more fertilizer and employ better machinery on their best land. Farm surpluses actually grew as a result.

THE 1950s AND 1960s SOIL BANK

During and just after World War II, participation in conservation programs reversed course as both farm prices and the demand for food increased. But after the war ended, demand was reduced and the surpluses grew once again. Farm bills in the late 1940s and early 1950s did little to help the situation. The farm bill of 1956 enacted the Soil Bank which eventually retired 29 million acres of cropland by converting it to natural resource uses in exchange for three to ten years of government payments.

The Soil Bank had two parts, the acreage reserve program (ARP) and the Conservation Reserve Program (CRP). The ARP discouraged farmers from planting over-abundant crops such as cotton, wheat, corn, tobacco, rice, and peanuts. The CRP paid farmers to retire cropland in three-, five-, or ten-year contracts to promote the quality of soils, water, forest, and wildlife (Helms 1985). The ARP functioned from 1956 to 1958 but ended for many of the same reasons that had caused portions of the ACP to fail in the 1930s. Surpluses continued to grow and critics of the program "contended that [it] was an expensive means for buying reduced production and higher farm incomes" (Helms 1985, 1).

ARP payments were higher than those under the CRP, which caused a lack of interest in CRP land retirement. Increases in funding after 1958 made the longer-term CRP land retirement options more attractive and by 1960 nearly 29 million acres had been enrolled in the program. Although contractual payments to owners of CRP-enrolled land continued after 1960, no new lands were enrolled in the program.

"FENCEROW-TO-FENCEROW" IN THE 1970s

Problems with surpluses persisted through the 1960s with crop carryovers at or near historic highs for corn, wheat, barley, and sorghum. Prices fell to levels not seen since the early 1940s. Some farmers spent their conservation payments on fertilizer and drainage improvements, both of which improved yields on remaining cropland.

Crop shortages around the world and a falling U.S. dollar led to increased export demand for American-grown crops. Exports doubled between 1972 and 1973, making it nearly impossible to justify supporting conservation programs designed to limit production (Bowers and others 1984). As soon as conservation contracts expired farmers plowed up grasslands to reestablish crop production. The few revisions to conservation programs that took place in 1970s farm bills dealt mostly with issues such as protecting waterfowl breeding grounds and addressing water quality issues related to pollution from livestock wastes. Much of what had been accomplished with cropland retirement over the preceding forty years was erased. Farmers planted crops on as much ground as they could acquire, whether through ownership or rental.

THE CONSERVATION RESERVE PROGRAM, 1985 TO 2007

Farm bills of the 1980s focused on conservation. The Food and Security Act of 1985 included programs such as Sodbuster (which protected highly erodible lands) and Swampbuster (which prohibited wetlands from being converted to cropland). The current Conservation Reserve Program (CRP) was established as part of the 1985 farm bill. It is a voluntary program consisting of 10- to 15-year contracts whereby landowners receive per-acre payments based on soil productivity and average dry-land cash-rent equivalents. The program awards cost-shares of up to 50% to establish long-term conservation land covers to protect environmentally sensitive land.

The original CRP of 1985 has been reauthorized and amended in subsequent farm bills, and each reauthorization has brought changes in how the program is administered (National Agricultural Law Center 2015). The initial goal of the CRP was to remove 40–45 million acres of cropland from production out of the nation's 445 million-acre total. The 1990 farm bill expanded the definition of "environmentally sensitive" lands to include additional threats to the environment, thus increasing acres eligible for enrollment. Not only were highly erodible lands eligible in the 1999 law, so also were marginal pasturelands and croplands that threatened water quality.

Between 1986 and 1994 the CRP enrolled 35 million acres of the Great Plains, southern Iowa, and the western fringe of the Corn Belt, northern Missouri, southern Idaho, and eastern Washington which formed the core of the program (Auch and others 2013). Enrollments were greatest from 1987 to 1990 when 30.6 million acres of the 35.0 million acre 1994 total entered the program (USDA FSA 2015). To keep payments from becoming a disproportionately large source of farm income in any single county—which could be to the detriment of businesses that rely upon farming and farmers—no county could enroll more than 25% of its total cropland acreage in CRP. The CRP was reauthorized in the 1996 farm bill. The enrollment cap was lowered to 36.4 million acres, but criteria for eligibility were broadened to encompass the retirement of lands deemed beneficial to wildlife (Figure 10.1).

The bill also created an option known as continuous sign-up. Prior to 1996 sign-ups were competitive. Landowners submitted bids to county Farm Service Agency offices and were chosen based on eligibility criteria. Continuous sign-ups were non-competitive and could be entered into at any time of the year. Typically, prior to 1996, entire fields (individual parcels of land) were retired, but continuous sign-ups only retired the most environmentally vulnerable areas within parcels, such as strip of land on either side of a stream. This left the farmer the option of continuing to farm the rest of the parcel. The more focused programs also paid landowners a higher rate per acre. The enrollment cap was increased to 39.2 million acres in 2002 and new continuous options, such as the Farmable Wetlands Program (FWP), were introduced.

Enrollment rebounded from 1999 to 2007 and culminated with 36.74 million acres, the highest in the program's history. CRP enrollment grew in the core region during this time, but interest began to decline elsewhere. Some farmers who participated during the CRP's formative years opted out

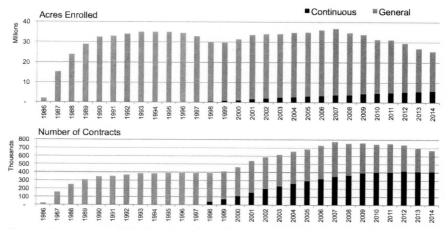

Figure 10.1   Acres enrolled and number of continuous and general contracts for the Conservation Reserve Program, 1986 to 2014. *Source*: Created by authors, data from USDA Farm Service Agency.

due to perceived or realized negative aspects of the program. High crop prices no doubt were a factor in returning lands to production (Laingen 2013).

CONSERVATION RESERVE PROGRAM AFTER 2007

The enrollment cap was reduced to 32 million acres in 2008 due to federal budget restrictions. More than 11 million acres were removed from the program between 2007 and 2014, and the number of general enrollment contracts dropped 40%. The year 2007 marked a turning point in participation. Counties that had large areas of land enrolled in CRP since the late 1980s had gone through a 10-year contract renewal period in the late 1990s. For them, 2007 was a second opportunity to opt out of the program. Passage of the Energy Policy Act of 2005 and the Energy Independence and Security Act (EISA) of 2007 mandated the production of 26 billion gallons of renewable fuel by 2022. Corn and soybean prices increased to the point that farmers could earn greater financial benefits from farming their land than they could keeping it retired in programs such as the CRP.

Since 2007, South Dakota has lost over 40% of its CRP acreage, dropping from just under 1.6 million acres to just over 900,000. On average, in the early 1990s South Dakota landowners willing to participate in the CRP could earn about $40.00 per acre in CRP rental payments compared to cropland rental payments of $32.00 per acre (Janssen and others 2015). Since 1991, per acre CRP payments have increased $0.92 per year, while cropland rental

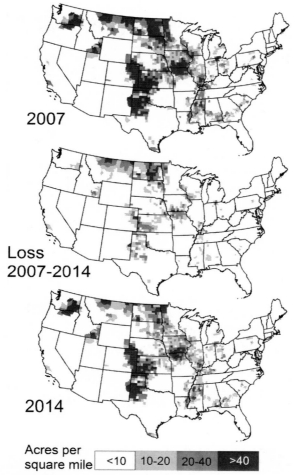

Figure 10.2   Acres per square mile of CRP in 2007 and 2014, and loss between the two years. *Source*: Created by authors, data from USDA Farm Service Agency.

prices increased at a rate of $4.37 per year. By 2014 CRP payments state-wide were $72.00 per acre and cropland rental prices were $150.00 per acre. In 23 years, someone who owned cropland and had the option of putting it into the CRP or renting it out for production went from making money to losing money if they kept the land in the CRP.

Regions that had notable CRP loss after 2007 included the eastern Dakotas and the High Plains of Kansas and Colorado, although there was a more general trend of CRP reduction across the entire country (Figure 10.2). CRP general contract acreage has declined nationwide but continuous sign-ups have increased. Although smaller in terms of acreage, their more targeted approach enhances the landowner's ability to protect and conserve natural resources.

The Agricultural Act of 2014 (2014 Farm Bill) further reduced the cap on lands enrolled in the CRP to 24 million acres by 2017. Continuous sign-ups are targeted for high-priority parcels such as riparian buffers, filter strips, grass waterways, and wetland restoration. Such applications provide higher per-acre benefits than do large-tract, whole-field land retirements (USDA ERS 2015). This flexibility has allowed farmers to focus on, and react to, what is best for their operations in terms of both production and conservation.

REFERENCES

Auch, R. F., C. R. Laingen, M. A. Drummond, K. L. Sayler, R. R. Reker, M. A. Bouchard, and J. J. Danielson. 2013. Land-Use and Land-Cover Change in Three Corn Belt Ecoregions: Similarities and Differences. *Focus on Geography* 56(4): 135–143.

Bowers, D. E., W. D. Rasmussen, and G. L. Baker. 1984. History of Agricultural Price-Support and Adjustment Programs, 1933–84. *Agricultural Information Bulletin 485*. Washington, DC: USDA ERS.

Cain, Z., and S. Lovejoy. 2004. History and Outlook for Farm Bill Conservation Programs. *Choices* 19(4): 37–42.

Gersmehl, P. J., and D.A. Brown. 2004. The Conservation Reserve Program: A Solution to the Problem of Agricultural Overproduction? In *WorldMinds: Geographical Perspectives on 100 Problems,* edited by D. G. Janelle, B. Warf, and K. Hansen. Boston: Kluwer.

Helms, J. D. 1985. Brief History of the USDA Soil Bank Program. *Historical Insights* 1, USDA, NRCS. http://www.nrcs.usda.gov/Internet/FSE_DOCUMENTS/stelprdb10455666.pdf.

Janssen, L., J. Davis, and S. Adams-Inkoom. 2015. *South Dakota Agricultural Land Market Trends 1991–2015.* iGrow Research. USDA-SDSU Agricultural Experiment Station, Publication 03-7008-2015.

Laingen, C. R. 2011. Historic and Contemporary Trends of the Conservation Reserve Program and Ring-Necked Pheasants in South Dakota. *Great Plains Research* 21: 95–103.

Laingen, C. R. 2013. A Geo-temporal Analysis of the Conservation Reserve Program: Net vs. Gross Change, 1986 to 2013. *Papers in Applied Geography* 36: 37–46.

Leathers, N., and L. M. B. Harrington. 2000. Effectiveness of Conservation Reserve Programs and Land "Slippage" in Southwestern Kansas. *The Professional Geographer* 52(1): 83–93.

National Agricultural Law Center. 2015. *United States Farm Bills.* http://nationalaglawcenter.org/farmbills/

Reganold, J. P., D. Jackson-Smith, S. S. Batie, R. R. Harwood, J. L. Kornegay, D. Bucks, C. B. Flora, J. C. Hanson, W. A. Jury, D. Meyer, A. Schumacher, Jr., H. Sehmsdorf, C. Shennan, L. A. Thrupp., and P. Willis. 2011. Transforming U.S. Agriculture. *Science* 322(6030): 670–671.

USDA ERS. 2015. *Agricultural Act of 2014: Highlights and Implications.* http://www.ers.usda.gov/agricultural-act-of-2014-highlights-and-implications/conserva-tion.aspx.

USDA FSA. 2012. *Conservation Reserve Program: Annual Summary and Enroll-ment Statistics FY 2012.* U.S. Department of Agriculture, Farm Service Agency, Washington, DC. http://www.fsa.usda.gov/Assets/USDA-FSA-Public/usdafiles/Conservation/PDF/summary12.pdf.

USDA FSA. 2015. *CRP Enrollment and Rental Payments by State, 1986–2014.* U.S. Department of Agriculture, Farm Service Agency, Washington, DC. http://www.fsa.usda.gov/Assets/USDA-FSA-Public/usdafiles/Conservation/Excel/statepymnts8614.xls.

Appendix

Keeping Track of Production

Government agencies have been collecting, disseminating, and archiving agricultural data for over 150 years. Data collected through censuses, surveys, aerial photography, and satellite imagery are used to keep track of production and to monitor how the land is being used. Three of the most frequently used agricultural datasets focused on U.S. farms, farmers, agricultural production, and land use are the quinquennial Census of Agriculture, the annual National Agricultural Statistics Service (NASS) June Area survey, and a raster-based land-cover data product called the Cropland Data Layer. Depending on the question that is being asked or the data being sought, the answer is likely to be found in one of these three collections of data.

CENSUS OF AGRICULTURE

The U.S. Census of Agriculture is the largest repository of agricultural data about the United States at the national, regional, state, and county scales (USDA 2015a). The most recent Census of Agriculture (2012) marked the twenty-eighth time that the government has collected statistics about agricultural products, economy, land use, and farm operators. The Census is the only uniform source of farm-related data for every state, county, or county equivalent in the nation. These data are free of charge to the user and are easily accessible either as bound copies or via the Internet: http://agcensus.usda.gov.

Congress first appropriated funds to investigate agricultural production and practices in 1839. This led to the first Census of Agriculture that was taken as part of the sixth 10-year Census of Population in 1840 (USDA 1969). The cooperative population and agricultural censuses continued until 1920, at which time the Census of Agriculture interval was changed to every five years.

This resulted in a mid-decade census being taken in 1925, 1935, and 1945. After the dual population/agricultural census was taken in 1950 the Census of Agriculture was conducted in years ending in –4 and –9. This changed in 1976 when the U.S. Congress authorized that it be administered in 1978 and 1982 and changed the 5-year collection cycle to years ending in –2 and –7 so that the Census of Agriculture coincided with other economic censuses that cover manufacturing, mining, construction, trade, and transportation.

The Appropriations Act of 1997 removed the responsibility of administering the Census of Agriculture from the Department of Commerce's Bureau of the Census and awarded it to the USDA's National Agricultural Statistics Service (NASS) where it remains today. While what defined a farm did not change, NASS reviewed mailing lists used for the five preceding censuses and discovered they had failed to include 8% of all U.S. farms in the 1997 Census (Hart and Lindberg 2014). Because of this oversight, the 1997 Census contains two sets of data: one comparable to earlier censuses and a second that is referred to as "adjusted for coverage." While this "adjustment" increased, overnight, the number of farms in the United States from 1.9 million to 2.2 million (+304,017), most were small farms that produced and earned very little. When using data from 1997 and subsequent censuses it is important to understand this distinction and use/report data accordingly.

Data collected for the Census of Agriculture originate with individual farms. Since the start of the Census the definition of a farm has changed nine times—most recently in 1974 when it became "any place from which $1,000 or more of agricultural products were produced and sold, or normally would have been sold, during the census year" (USDA 2015a, A-1). All data are subjected to disclosure review prior to publication to protect respondent confidentiality. If data reported by the Census would allow someone to determine a particular farmer's identity, those data are suppressed and coded with a "D" indicating nondisclosure. This typically occurs in counties where there are very few farmers engaged in a particular kind of farming activity.

Data are aggregated from the individual farm scale to the county, state, and national levels. This hierarchical system is employed throughout the Census and allows for analysis at multiple scales, which is especially helpful for geographic applications. Definitions are nested to make data easier to aggregate.

For example, the category "Land in Farms" includes lands used for crops, pasture, and grazing as well as other lands not under cultivation such as woodland, wasteland, or lands in various conservation programs. Cropland, and other subsets of Land in Farms, are further broken down into more exact types of uses. Total Cropland comprises lands where crops were harvested, pasture or grazing lands that could have been used for growing crops, other cropland such as idle cropland or cropland on which crops failed, and cropland that was left fallow. The hierarchy continues with individual crops reported by acres harvested which, once summed, account for acreages under the Harvested Cropland category. Quantities of individual types of crops and livestock products such as bushels of corn, gallons of milk, and tons of hay, to name a few, are also reported.

Data from the 2002, 2007, and 2012 censuses can be downloaded for use in database or Geographic Information Systems (GIS) software using the Desktop Data Query Tool. For the 2012 census, the Desktop Data Query Tool can be found at: http://www.agcensus.usda.gov/Publications/2012/Online_Resources/Desktop_Application/. Users can download an executable (.exe) file, install it on their PC, use the software to query data at the state and county scale, and export user-defined data as a database (.dbf) file or a comma separated values (.csv) file. Digital data stored on CDs from pre-2002 censuses are available upon request to the USDA but are subject to availability.

The most reliable source for historical census data (other than actual bound hardcopies) are the scanned documents available from the National Agricultural Statistics Service via the Albert R. Mann Library at Cornell University in Ithaca, New York. These documents have been seamlessly integrated into the Census of Agriculture's web interface. Data availability at the state and county scales, which form the majority of categories present in more contemporary censuses, is quite good dating back to 1925, the year many consider to be the first modern census of agriculture, and whose categories and format have remained largely unchanged to today.

As with any census or survey a certain level of error is inherent in the data. Total coverage is impossible to achieve. According to the USDA, the 2012 Census had a response rate of 80.1%, down from 85.2% in 2007 to 88.0% in 2002 (USDA 2015a). NASS maintains a list of places that meet the definition of a farm from which their Census Mailing List (CML) is compiled. NASS endeavors to keep this list current and makes adjustments to it as new data are acquired. While the process of collecting data for the Census is not inherently sample-based, the Census does use sampling procedures in compiling the CML and in its data collection procedures such as when data are processed and edited.

Bias, nonresponse, undercoverage, and misclassification are compensated for through the use of weighted adjustments based on statistical measures (For a full explanation of the methods used to collect and compile the Census of Agriculture, see Appendix A—Census of Agriculture Methodology [USDA 2015a]. The census forms that respondents fill out and return are displayed in Appendix B—General Explanation and Report Form [USDA 2015b]).

The USDA has a long history of producing maps of data collected from the Census. Access to maps can be found by using the drop-down menu where individual census years can be selected. For recent censuses, this product has been referred to as the "Agricultural Atlas." For censuses from 1982 through the most recent, links can be found within those pages that refer to a "Graphic Summary" or special reports that include a link to "Maps." As a user browses the USDA's online census archives (or bound hardcopies) more frequently, the better-acquainted one becomes with where data, maps, and other information are found. Volumes from different years may be organized in a slightly different manner, but before long users deduce trends in where data—especially obscure, less frequently queried categories—can be found.

NASS ANNUAL STATISTICS

Each June NASS conducts one of its largest annual surveys. Data gathered through the June Area survey are the foundation on which the NASS survey program functions and they provide statistical measures for quality assurance involving other NASS products (USDA, NASS 2015). The framework for the survey consists of nearly 11,000 one square-mile parcels of land from which land-use data are obtained. Farmers who operate individual land units within each square mile surveyed are interviewed. NASS reports that most years approximately 85,000 individual tracts of land from within the 11,000 sample areas are identified. These produce some 35,000 interviews conducted with individuals responsible for farming or maintaining that land (USDA, NASS 2015). More details on the June Area survey as well as other NASS data products can be found at http://www.nass.usda.gov/Surveys/ Guide_to_NASS_Surveys/.

Data collected by the June Area survey become available for statistical and spatial analyses through the NASS Quick Stats interface: http://quickstats. nass.usda.gov. Here, users can choose items from selection boxes and query annual data. Searches begin with users selecting their variable of interest, a measure or units value (e.g. area, tons, yield), the spatial scale of the data (county, district, state, or national), and the temporal span (a single year or multiple years) needed. After the query is performed users have options to save their search, download the data as a .csv file, print the data, or—if the data queried allow—create a map. Data downloaded at the state or county scale include Federal Information Processing Standards (FIPS) codes to use as unique identifiers for use in joining to spatial data in a GIS for custom mapping.

Data queried using Quick Stats, like all data compiled from the June Area survey, are derived using a sampling framework. Users may note that returns for identical categories from the Census of Agriculture and the June Area survey differ for the years where both the Census and June Area survey occur. These discrepancies are caused by differences in the methods used for compiling these two datasets. For example, the Census of Agriculture reported that farmers harvested 87,413,045 acres of corn for grain in 2012. The June Area survey reports 87,365,000 acres. At a more local scale, using corn for grain in 2012, the Census reported McLean County, Illinois, as having harvested 347,414 acres, whereas the June Area survey reports 341,000 acres. Mixing June survey data with Census data should be avoided under most circumstances.

CROPLAND DATA LAYER

The Cropland Data Layer (CDL) is a raster-based data product created and released annually by the USDA's National Agricultural Statistical Service. Information about the CDL can be found at: http://www.nass.usda/gov/Research_and_Science/Cropland/SARS1a.php. In the 1970s and early 1980s NASS had a goal to use multi-spectral satellite imagery to estimate area of major commodity crops, including corn, soybeans, and wheat (Craig 2010). In current usage, data derived from the CDL provide crop-acreage estimates at the state and county levels. Data offer background information for the NASS June-based acreage survey and are used to identify cropland areas impacted by disasters. CDL and other remotely sensed data products are checked with ground surveys ("ground truth") for accuracy (Johnson 2013).

Inputs into the current CDL data come from five sources: (1) the AWiFS (56-meter resolution) sensor mounted on a satellite flown by India; (2) NASA Thematic Mapper (TM) and Enhanced Thematic Mapper (ETM+) imagery at 30-meter resolution via Landsat 5, 7, and 8 satellites; (3) Terra's MODIS (250-meter resolution) satellite; (4) the USDA Farm Service Agency's (FSA) Common Land Unit (CLU) data used for ground truth; and (5) National Land Cover Database (NLCD) data that are used to classify image-elements deemed "non-agricultural." For a more detailed treatment on hardware and data specifications and methods, see Boryan and others (2011) and Johnson and Mueller (2010).

Data can be interactively assessed and downloaded using the CDL's web-interface known as CropScape (http://nassgeodata.gmu.edu/CropScape/). CropScape is a geospatial data service that began in 2011 as a collaborative effort between NASS and the Center for Spatial Information Science and Systems at George Mason University. CDL data are available via CropScape from 1997 (when North Dakota was the only state with data available) to the present. Additional states were subsequently added, with 2008 marking the first year of coverage for all of the Lower 48 states. Users seeking national-scale datasets for use with their own GIS software applications can download national CDL zip files. Beginning in 2008, most of these files were between one and two gigabytes in size. Data for the most current year are typically uploaded by spring of the following year (2015 national CDL data were released February 12, 2016).

CropScape's interactive features include zooming, panning, defining areas of interest, data export, geospatial queries, land-use change analysis, map creation (in PDF format), automated delivery of data to services such as Google Earth, and the ability to create tabular or graphical reports of crop types and amounts found nationally or in states, counties, or user-defined areas of interest. The online interface includes a robust "Help" section, as well as a demonstration video for new users and a link to the frequently asked questions page.

Data originating from the CDL embrace more than 130 classes of agricultural and non-agricultural land cover. Because these data are a product of classifying remotely sensed imagery, they contain an inherent amount of error, as do any product that classifies vegetative land covers into specific categories (Laingen 2015). Widespread and easily identifiable (spectrally unique) crop types such as corn, soybeans, or cotton typically have high accuracy rates (more than 85%). Discriminating between less common crops or crops whose spectral signatures closely resemble one another is more likely to produce error. Confusing small grains with non-agricultural grasses is another example of error.

The CDL also includes spectrally ambiguous land-cover classes such as alfalfa and non-alfalfa hay and fallow versus idle cropland. These cover types typically have much lower accuracy rates—in some cases less than 60%, and in other cases even lower. Error matrices and accuracy assessments for CDL data are available at the state scale by crop type under the Metadata link on their homepage: (http://www.nass.usda.gov/Research_and_Science/Cropland/metadata/meta.php).

OTHER SOURCES OF INFORMATION

1. ERS Charts of Note: http://www.ers.usda.gov/data-products/charts-of-note.aspx.

    The USDA's Economic Research Service (ERS) generates many useful types of charts and graphs. Users may provide an email address and have alerts sent to them directly when new items are released. Typically the charts consist of interesting or timely maps, illustrations, and other graphics related in some way to agriculture that have been selected by ERS staff from publicly released ERS reports. Topics include nutrition, bioenergy, food safety, agricultural economics, rural poverty, global agricultural trade, and many others. Links are provided to the more detailed reports from which the charts were taken. All of the charts can be found at: http://www.ers.usda.gov/publications.aspx.

2. FAOSTAT: http://faostat3.fao.org/home/e.

   The Food and Agriculture Organization (FAO) of the United Nations collects and disseminates data on agricultural production, trade, land use, food and food security, prices, and other associated topics for hundreds of participating countries in the world. User-friendly databases and interfaces are accessed via the organization's webpage, where data can be easily searched and downloaded. Other FAO statistics available for most of the world's countries can be accessed in the same way for fisheries (AQUA-STAT), and for forestry and forest products (FORESTAT).

3. EROS Remotely Sensed Data and Pre-Packaged Products: http://eros. usgs.gov.

   Satellites orbiting the Earth have been collecting land-cover data at multiple spatial resolutions since the early 1970s. Aerial photographs used for agricultural assessments have been used for nearly a century. Raw satellite and aerial imagery and ready-to-use land-cover data can be easily obtained through data download portals made available by the U.S. Geological Survey's Earth Resources Observation and Science (EROS) Center. USDA Farm Service Agency aerial photography is available there or through the USDA's Geospatial Data Gateway: https://gdg.sc.egov.usda.gov.

4. Conservation Reserve Program data: http://www.fsa.usda.gov.

   Nearly thirty years of monthly, annual, and historical data, as well as general information about the Conservation Reserve Program and other government-funded agricultural-centric conservation initiatives, are available at the main USDA Farm Service Agency webpage. Users should click on Programs and Services and select Conservation Programs. Detailed monthly and annual reports are available in PDF format. Historical data at the state and county scales are available via downloadable Excel tables.

REFERENCES

Boryan, C., Z. Yang, R. Mueller, and M. Craig. 2011. Monitoring U.S. Agriculture: the U.S. Department of Agriculture, National Agricultural Statistics Service, Cropland Data Layer Program. *Geocarto International* 26(5): 341–358.

Craig, M. 2010. *A History of the Cropland Data Layer at NASS.* http://www. nass.usda.gov/Research_and_Science/Cropland/CDL_History_MEC.pdf. USDA-NASS, Washington, DC.

Hart, J. F., and M. B. Lindberg. 2014. Kilofarms in the Agricultural Heartland. *Geographical Review* 104(2): 139–152.

Johnson, D. M. 2013. A 2010 Map Estimate of Annually Tilled Cropland Within the Conterminous United States. *Agricultural Systems* 114: 95–105.

Johnson, D. M., and R. Mueller. 2010. The 2009 Cropland Data Layer. *Photogrammetric Engineering and Remote Sensing* 76(11): 1201–1205.

Laingen, C. R. 2015. Measuring Cropland Change: A Cautionary Tale. *Papers in Applied Geography* 1(1): 65–72.

U.S. Department of Agriculture. 1969. *The Story of U.S. Agricultural Estimates.* Statistical Reporting Service, Miscellaneous Publication No. 1088. http://www.nass.usda.gov/About_NASS/The%20Story%20of%20U.S.%20Agricultural%20Estimates.pdf. USDA-NASS, Washington, DC.

U.S. Department of Agriculture. 2015a. *2012 Census of Agriculture, Appendix A.* http://www.agcensus.usda.gov/Publications/2012/Full_Report/Volume_1,_Chapter_1_US/usappxa.pdf. USDA-NASS, Washington, DC.

U.S. Department of Agriculture. 2015b. *2012 Census of Agriculture, Appendix B.* http://www.agcensus.usda.gov/Publications/2012/Full_Report/Volume_1,_Chapter_1_US/usappxb.pdf. USDA-NASS, Washington, DC.

USDA, NASS. 2015. *June Area.* http://www.nass.usda.gov/Surveys/Guide_to_NASS_Surveys/June_Area/. USDA-NASS, Washington, DC.

INDEX

About the Authors

John C. Hudson is Professor and Director of the Program in Geography at Northwestern University in Evanston, Illinois. His research, which has focused on the American Midwest, has been published widely in professional journals and books. He is the author of *Plains Country Towns* (1985), *Making the Corn Belt* (1994), *Across This Land: A Regional Geography of the United States and Canada* (2002), and *Chicago: A Geography of the City and Its Region* (2006).

Christopher R. Laingen is Associate Professor of Geography at Eastern Illinois University in Charleston, Illinois. He grew up on a family farm in southern Minnesota near the town of Odin, where much of his interest in the rural and agricultural geography began. His research focuses on changes in farming and the landscapes of the rural Midwest and Great Plains, which has been published in *Great Plains Research, Focus on Geography, The Geographical Review,* and *The Professional Geographer.*